261398 07 1

UNDERSTANDING ELECTRONIC COMPONENTS

UNDERSTANDING ELECTRONIC COMPONENTS

IAN R. SINCLAIR
B.Sc., M.I.E.E., M.Inst.P

FOUNTAIN PRESS: LONDON

Fountain Press
M.A.P. Book Division,
Station Road,
Kings Langley,
Hertfordshire

First Published 1972

ISBN 0 852 42104 4

Set in Monophoto Times and printed in England by
Page Bros (Norwich) Ltd., Norwich

CONTENTS

PREFACE

EVERY BOOK DEALING WITH ELECTRONICS must, unless it is to take up more shelf space than an encyclopaedia, assume some existing knowledge on the part of the reader and leave him with more to learn from further texts. This book is no different in that it assumes some experience in the use of electronic components, and leaves much of the circuit side of electronics to the companion volume *Understanding Electronic Circuits.*

Previous knowledge which is assumed is the use of Ohm's Law and the use of the units of voltage and current. Some degree of practical experience, perhaps in audio, radio or TV circuits is also assumed, so that this is not a book for the absolute beginner but rather for those who have entered the second stage of interest in electronics where the surprise of seeing something work is replaced by curiosity about *how* it works.

For some time, there has been a scarcity of texts directed to this stage of interest. Most previous books have either assumed no previous knowledge at all, so concentrating on the very elementary parts of electronic circuits with scant mention of actual components, or they have been aimed at the engineer, assuming a good background of experience and developing the subject to a high theoretical level.

This book is aimed towards the amateur who wishes to know more about the components he uses, to the technician engineer who finds some of the standard reference works on components rather unapproachable, and to the service engineer wishing to use the book as a guide to his work, or as a text for a course of study. Students of electronic engineering should also find this a useful reference book in their first year of study, and design engineers will find it a useful summary of much information on electronic components which would otherwise be available only in a scattered form. In this context, the book may find considerable use as a reference work in the purchasing departments of firms concerned with buying electronic components, and also in the sales departments of firms engaged in the distribution of such components.

Such a work as this can be produced only with the wholehearted co-operation of the electronics industry and I wish to thank those who gave up their valuable time to discuss the subject of electronic components with me. I also wish to acknowledge the help, information and other assistance which have been unstintingly given by the many firms approached during the writing of this book. In particular, the following merit special mention:

The Belclere Co. Ltd., Belling and Lee Ltd., B & R Relays Ltd., Colvern Ltd., Decca Radar Ltd., RS Components Ltd., Gardners Transformers Ltd., Jermyn Industries Ltd., Miniature Electric Components Ltd., Morganite Ltd., Mullard Ltd., Plessey Ltd., S.T.C. Ltd., Welwyn Electric Ltd., and the companies forming the GEC Group.

I should also like to record the contribution of my wife, without whose encouragement and continual assistance this book would not have been written.

I. R. Sinclair *January, 1972*

CHAPTER ONE

INTRODUCTION

ALL ELECTRICAL AND ELECTRONIC equipment depends on effects produced by the flow of units of electricity called electrons. For our purposes, we can regard electrons as particles of unimaginably small size, each repelling any other because of the electrical charge which each carries. By convention, we regard this charge which causes the repulsive force as being *negative* for the electron. A charge is said to be *positive* if it causes electrons to be attracted, and the size of the charge is measured by the force which it exerts on another charge.

Electric Current

Materials which conduct electricity well do so because they contain large numbers of electrons which are free to move; conversely, insulators contain very few electrons free to move. It is the movement of the electrons (or other charged particles in some cases) which constitutes electric current, and the number of electrons passing a given point in the time of one second is a measure of the strength of electric current. In a good conductor, where huge numbers of electrons move, the average speed of the electrons is very low, perhaps only a few cm per hour; in a poor conductor carrying the same current the average electron speed would have to be much higher, perhaps as much as ten million metres per second in a vacuum.

Since electric current is the movement of charged particles such as electrons, we can speak of current as flowing when the electrons move to-and-fro in the conductor, just as current flows when the electrons move in one direction. Before anything was known of electrons, these two types of current were recognised and given the names a.c., meaning alternating, or oscillating, current, and d.c., or direct current. In electronics, a.c. is vastly more important, since all electronic signals consist of a.c., and d.c. is an incidental, used mainly for power supplies.

Electrical Quantities

The three quantities of most interest to us are Voltage, Current and Frequency. Current we have met already; it is a measure of the number of electrons passing any point in one second. The voltage of any point is a comparative figure; it measures the tendency of electrons to move to or from that point to another one. By convention, we usually take the voltage of the Earth's surface as that 'other' voltage, so that we talk of a point being some number of volts above or below **earth**. When the voltage at a point (compared to earth) or the current in a circuit reverses at regular intervals, we are dealing with an alternating voltage or current, and the number of such reversals per second in one direction (for example, positive to negative) is called the frequency.

The frequency of the a.c. power supply in this country is 50Hz, the Hertz (abbreviated to Hz) being the unit of frequency represented by one complete cycle per second. When d.c. is present, it would be more conveniently referred to as z.f., meaning zero frequency of current or voltage, a style which avoids such nonsense phrases as 'd.c. voltage'. However, since this terminology has not been widely adopted at this point of time, and its use may lead to confusion, the conventional (but more inaccurate) term 'd.c.' will also be used.

Active and Passive Components

Electrical circuits consist of complete circular paths for current, starting at one end of a source of voltage (e.m.f.) and ending at the other end of the source. Inside the source, which may be a battery, an alternator or other device, energy of some sort is being used to push electrons round the circuit. Electronic circuits are built up by connecting components together by wire or other conducting material, or through the effects of electrostatics or magnetism. Components generally are of two types, called **Active** or **Passive**, according to whether or not they can be made to increase the power of a signal applied to them. This definition is not entirely watertight, but is as near as we can get without becoming too elaborate.

A passive component can be used to reduce the signal power, to select a part of a signal by its voltage, its frequency or spacing in time from another signal, to change the shape of a waveform or to couple active components together; in every case the power of the signal is either decreased or unchanged, never increased. Resistors, capacitors and inductors are typical passive components; there is a fourth one called a gyrator which is encountered mainly in microwave circuits.

An active component can increase the power of a signal, and must be supplied with both the signal and a source of power to do so. In many familiar active devices the source of power is a steady voltage, the d.c. supply, and the signal is fed in at one part of the device and taken out at another. Valves and transistors are of this type, but there are active devices in which the input and output occur at the same place and must be separated by other means, as, for example, in tunnel diodes and parametric diodes. There are also active devices in which the power supply is a.c. and the signal is d.c. or very low frequency a.c.; one example is the magnetic amplifier.

In integrated circuits, both the active and the passive parts of the circuit are basically of the same material and are formed together within one block of semiconducting material. Since no part of the circuit is accessible, though there may be terminals at which passive components may be attached, the integrated circuit must be treated as a single active component which carries out a large number of operations upon a signal or signals applied to it.

Thin film circuits use the techniques of integrated circuit manufacture to form passive components on a miniature circuit. They may be used with transistors and diodes in the form of miniature 'chips' which are welded into the circuit; alternatively, the thin film circuit may contain the passive components for a number of integrated circuits.

Passive Components

The common purely passive components are resistors, capacitors and inductors. The factor which they have in common is that they obstruct the flow of alternating current (though they do so in different ways) and are therefore said to present an impedance to the circuit. Resistors have a form of impedance called resistance; they impede the flow of current equally at all frequencies from zero upwards (within limits, as we shall see later); in addition, the temperature of resistors increases when current flows through them, so that energy is lost from them in the form of heat. This means that any signal current passing through a resistor must lose energy (and power, since power is the rate of loss of energy).

Capacitors and inductors have reactances which vary with the frequency of the current passing through them. Capacitors have a very high reactance, amounting to complete insulation, to d.c. currents, and the reactance decreases as the frequency of signal applied is increased. Inductors have a very low reactance, almost zero in some cases, to d.c. but the reactance becomes greater as the frequency of the signal is increased.

When current passes through a reactance of either type, no power is converted into heat and there is no loss. In practice no reactance is perfect, particularly that of inductors, and each behaves as if it had some resistance included. We can measure the proportion of power-wasting resistance in such components; it is expressed as the *Power Factor* (see later).

Impedance is made up of resistance and reactance, and represents the total effect of resistance and reactance to a current. We cannot find impedance by simply adding resistance to reactance, however. We have to find impedance in the same way as we find the distance between two places on a map, using resistance and reactance as the map co-ordinates or references. We can increase the similarity of the problem by mapping resistance, reactance and impedance on a diagram called a phasor diagram (Fig. 1.1), where resistance is always measured off in one direction and reactance measured in a direction at right angles to the direction of resistance.

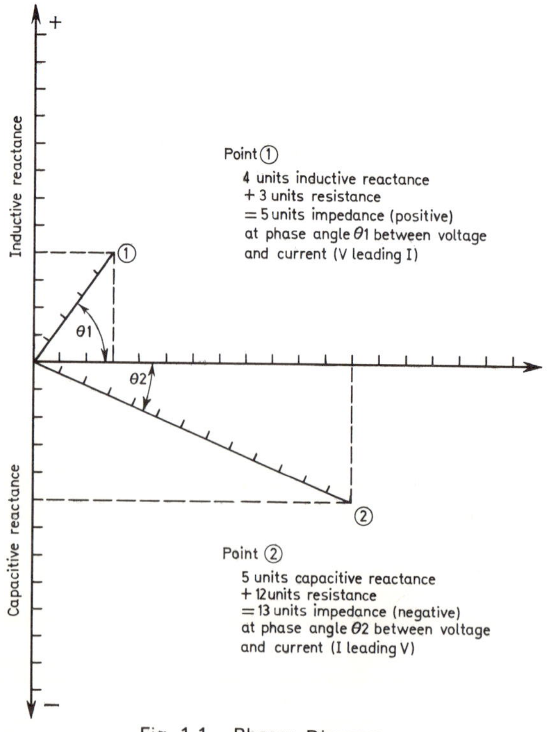

Fig. 1.1 Phasor Diagram

Why at right angles? If we connect a resistance and a reactance in series and pass alternating current through both of them, then the oscilloscope shows us something undetectable by meters. If we use an oscilloscope capable of displaying two waveforms simultaneously (a double beam 'scope), and we use one trace for the voltage across the resistor and the other for the voltage across the reactance, the waveforms appear to be out of step in time. The amount of the displacement is a quarter of a cycle, meaning that one waveform is reaching its peak just as the other passes through the zero voltage line (Fig. 1.2).

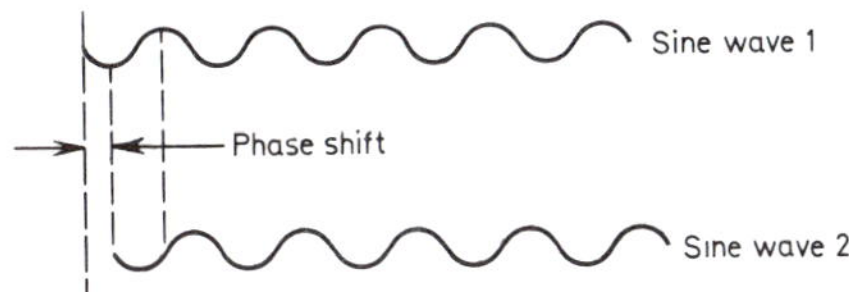

Fig. 1.2 90° Phase Shift.

When any event takes place in a cycle, this means that it starts and ends one cycle at the same value (of voltage, current, position, size or whatever is being cycled), and we can represent a complete cycle by a circle. The angle which a radius of a circle turns through so that the end sweeps out a complete circle is 360°, and a quarter of this is 90°. For this reason, the displacement of a quarter of a cycle is called a 90° phase shift, and this is the reason for talking about a phasor diagram and drawing the direction of a reactance at right angles to the direction of resistance (see Fig. 1.3).

What is actually 90° out of phase is the current in the circuit compared to its voltage. The voltage across the resistor has been used as a measure of the current flowing through it, because a resistor

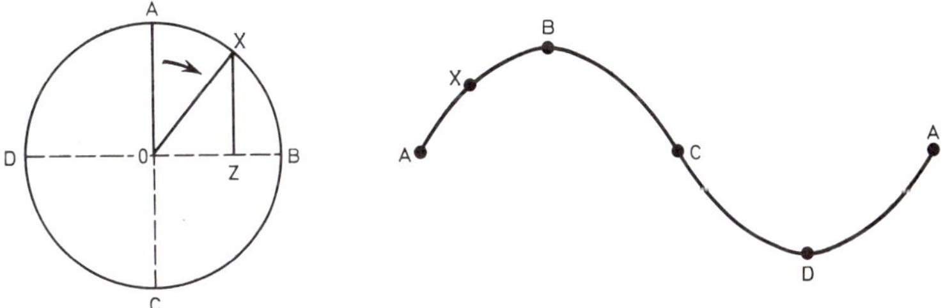

Fig. 1.3 The Circle and the Sinewave. If we imagine a line, OX, rotating as shown, then the length of its 'shadow', OZ, represents the amplitude of the sinewave. We can therefore label the points of the sinewave in terms of the angle through which the line was turned.

does not shift the phase of a current. Reactance does, however, shift the phase of current compared to voltage, and we have to draw the direction of reactance at 90° to resistance to show the correct size and direction of impedance. Note that the impedance as a whole must have some phase angle between volts and current.

The semiconductor diode can also be regarded as a passive component. It passes current in one direction only (the symbol used indicates a flow from positive to negative in the direction of the arrow) and so can be used to obtain a d.c. voltage from an alternating one. This type of use is termed *Rectification*. We can also use the diode as a coding device so that one type of signal passes and another is blocked. Uses of diodes will be considered in greater detail later.

Hardware

All the components which we have to use must be mounted in some way to allow the connections between them to be made. Equipment which uses large valves presents a sizeable mechanical problem, as the valve cooling arrangements will require either water circulation or the use of a cooling fan, and the weight of the valve will call for strong mounting which may have to be insulated, as many valve mountings are part of the anode or grid structures.

Smaller equipment almost universally uses printed circuit boards. An insulated board of laminated plastics forms the basis of the circuit, the material being chosen according to the needs of the circuit. For most purposes, the paxolin-type board is adequate, but boards using fibreglass, or with silicone resin plastics are found in specialised equipment where exceptional electrical insulation or heat resistance are required.

The boards are clad with a sheet of copper which is then etched away selectively to leave an outline of strips which form the circuit. Holes drilled at the end of the strips permit the wire ends of components to be lead through and soldered to the circuit strips. Heavier components may have tags attached which are then soldered to isolated pieces of copper. This form of mounting is completely satisfactory for all except those components which are either very heavy, run very hot, or have high voltages on the mounted terminals.

Failure of a printed circuit board (PCB) may occur through cracking of a circuit strip, usually caused by the board having been bent at some time. This is easily repaired by soldering over the crack, but care should be taken over re-mounting the board to avoid further cracking, and it is advisable to inspect for other cracks before replacing the board. Excessive heating of a mounted component may also cause the copper coating to become detached from board, so allowing the component to move and eventually break off, or faulty soldering may result in all contact being lost between the component and the board.

Printed circuit boards are not always the most desirable form of mounting for small electronic components. The facts of geometry often prevent simple, direct connections from being made and can lead to trouble in circuits operating at high frequency. This may be overcome by taking only d.c. supplies through the board and leading signals in and out separately by screened leads, or by mounting high frequency stages on special boards arranged to allow short connections to critical points.

Short-run and experimental circuits may be mounted on board such as *Veroboard*, which uses copper strips drilled at intervals (typically 0·15in) to take component leads. The strips may be cut and cross-connected at will to allow easy assembly of the circuit and easy change of components.

Wire used for linking between PCBs is usually insulated stranded flexible wire, preferably colour coded for easy identification. Short lengths of single strand, insulated or not, may be used for bridging between sections on a board.

Electrical Laws

The electrical laws of passive components most important for servicing work are:

(1) *Ohm's Law*—connecting impedance (measured in ohms) with voltage (in volts) and current (in amperes).—see Fig. 1.4

(2) *Kirchoff's First Law*—which states simply that the current leaving a circuit junction equals the current entering it. (Fig. 1.5)

(3) The Law of addition of resistors and reactors in series and parallel (for details see Fig. 1.6).

(4) The values of reactance of capacitors and inductors, and the way in which these vary with frequency (Table 1.1)

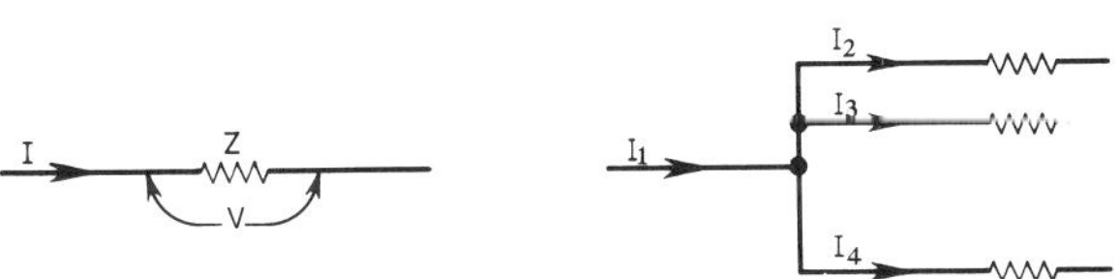

Fig. 1.4 (left) Ohm's Law. When a current (I) flows through an impedance of Z ohms, then the voltage measured across the impedance is Z × I volts.

Fig. 1.5 (right) Kirchoff's First Law. The current entering the junction equals the sum of the currents leaving the junction. $I_1 = I_2 + I_3 + I_4$.

Fig. 1.6 Adding resistors and reactors. (a) in series—the total resistance or reactance is the sum of the individual resistances or reactances, i.e. R1 + R2 = Total Resistance. (b) in parallel—the reciprocal resistances or reactances are added to give the total reciprocal resistance or reactance, i.e.

$$\frac{1}{R1} + \frac{1}{R2} = \frac{1}{\text{Total R}}$$

(5) The laws connecting watts dissipated (the energy given out) with voltage, current and resistance (Fig. 1.7)

Other Components

Chemists have known for a considerable time that electricity flow can take place in solution by the movement of positive particles (in the opposite direction to the negative ones) as well as negative particles, since the conduction of electricity through solutions depends on this. The resistance of such a solution depends on the total number of positive *and* negative particles. Electronic components using this principle have been used for integration (adding up the effect of a signal over a period of time).

Such a cell might consist of silver rods immersed in a solution of silver nitrate. When current flows, the effect is that the silver is dissolved from the positive rod and is plated on to the negative rod. If a form of measuring scale is included, this can be used as a crude measurement of time, as, for example, to measure roughly the time for which a piece of equipment has been switched on. It can also register other circuit times, as for example the total time for which a voltage or current has exceeded some preset value (the cell is in this case used with a gating circuit of some sort.)

In semiconductors, the flow of electricity depends also on movement of positive charges (called 'holes'). The action of transistors could not be accounted for unless it was assumed that current was carried both by electrons and holes.

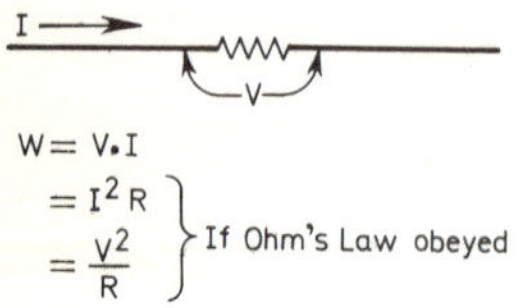

$$W = V \cdot I$$
$$= I^2 R$$
$$= \frac{V^2}{R}$$

If Ohm's Law obeyed

Fig. 1.7 Watts dissipated across a resistor.

Several passive components exist which perform functions which could hardly be carried out by resistors, capacitors and inductors. Such components are quartz crystals, whose impedance changes violently at one particular frequency; and a wide range of couplers and isolators which affect microwave signals in waveguide circuits.

f	L 10H	1·0H	0·1H	100mH	10mH	1mH	100μH	1μH
10Hz	630Ω	63Ω	6·3Ω	0·63Ω				
100Hz	6·3kΩ	630Ω	63Ω	6·3Ω	0·63Ω	negligible values		
1kHz	63kΩ	6·3kΩ	630Ω	63Ω	6·3Ω	0·63Ω		
10kHz	630kΩ	63kΩ	6·3kΩ	630Ω	63Ω	6·3Ω	0·63Ω	
100kHz			63kΩ	6·3kΩ	630Ω	63Ω	6·3Ω	0·63Ω
1MHz	Impractical values			63kΩ	6·3kΩ	630Ω	63Ω	6·3Ω
10MHz					63kΩ	6·3kΩ	630Ω	63Ω

REACTANCE OF INDUCTORS

f	C 10μF	1·0μF	100nF	10nF	1·0nF	100pF	10pF
10Hz	1·6kΩ	16kΩ	160kΩ	1·6MΩ	16MΩ	160MΩ	
100Hz	160Ω	1·6kΩ	16kΩ	160kΩ	1·6MΩ	16MΩ	160MΩ
1kHz	16Ω	160Ω	1·6kΩ	16kΩ	160kΩ	1·6MΩ	16MΩ
10kHz	1·6Ω	16Ω	160Ω	1·6kΩ	16kΩ	160kΩ	1·6MΩ
100kHz		1·6Ω	16Ω	160Ω	1·6kΩ	16kΩ	160kΩ
1MHz	Impractical		1·6Ω	16Ω	160Ω	1·6kΩ	16kΩ
10MHz	values			1·6Ω	16Ω	160Ω	1·6kΩ

REACTANCE OF CAPACITORS

Reactance of inductor = $2\pi fL$ where f is frequency in Hz, L is the inductance in H; 2π can usually be approximated by 6·3
Reactance of capacitor = $1/(2\pi fC)$, where C is the capacitance in farads

Table 1.1 REACTANCE FORMULAE

Basic Meters

Practically all the uses for meters in electronics work are met by some form of moving coil meter. Basically, such meters consist of a magnet with pole pieces shaped so as to concentrate the field to a central point, and a coil of wire rotating round this centre. The coil is supported by springs or by fine metal threads which have the task of supporting the coil and the pointer attached to it, setting the pointer to the zero of the scale and leading the current to and from the coil.

When current flows, the coil is magnetised and the magnet acts on it so as to turn the coil against the force of the springs. Where the magnetic force caused by the current equals the spring force, the coil and pointer come to rest, indicating a value on the scale. Such instruments can be made with very sensitive movements (1μA full scale reading), but most of the meters used have rather lower sensitivities ranging from 50μA full scale to about 1mA full scale.

Moving coil meters measure current; and any one instrument coil has its value of f.s.d. (full scale deflection) printed on it, measuring the current for which the meter pointer is at the maximum of the scale. Such a meter can be adapted to measure other ranges of current, and also to measure voltage. Other current ranges are measured by adding *shunts*. A shunt is a resistor wired in parallel with the meter so that it takes a known fraction of the current. To adapt the meter for voltage ranges, a resistor is added in series. The methods of finding the values of shunt and series resistors are shown in Fig. 1.8. When such a meter is arranged so that series and shunt resistors can be switched in, we have a Universal (or Multi-range) Meter.

Universal Meters

The Universal Meter, measuring a range of currents, voltages and resistances, is an indispensible tool for the service or design engineer. Unfortunately, the very usefulness of the meter often leads to its being used with insufficient thought—some care is necessary if the readings obtained are to be useful.

The most important basic fact is that all universal meters (as distinct from valve or field-effect transistor voltmeters) are basically *current measuring* instruments, which means that they take current from the circuit being measured, This current may be very small on the z.f. (d.c.) ranges, only 50μA being drawn when a Mk. VIII Avometer is reading full scale deflection. Current is being drawn, nevertheless, and this must be remembered when the instrument is being used for other purposes.

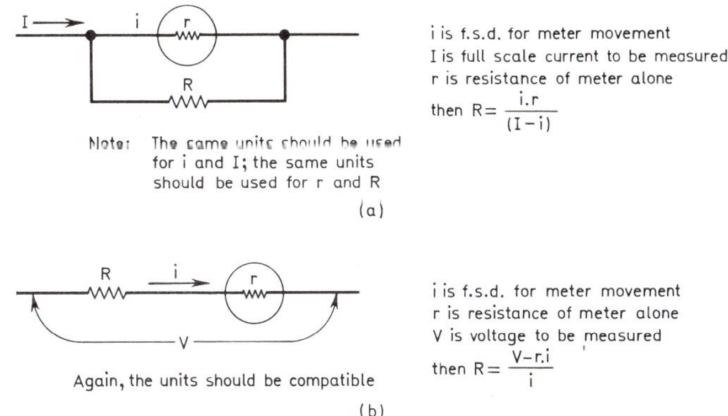

Fig. 1.8 (a) shunting a meter for a higher current range, (b) using a series resistor for a voltage range.

Measuring Current (z.f.)

If a meter is to be used to measure current flowing in a circuit, then either (i) the circuit must be broken so that the meter can be joined in series, set to the highest current range, the circuit switched on and the meter range switch altered until a reading of $\frac{1}{4}$ to full scale deflection obtained; or (ii) the *voltage* can be measured across a small (1Ω or so) resistor in series with the circuit, and the current calculated from Ohm's Law ($I = V/R$).

The first method is more accurate but requires the circuit to be broken. In some cases a clip-link may be provided in the circuit so that current readings may easily be taken, and in other circuits the breaking of the circuit may be easily carried out. Where high voltage pulses or r.f. voltages are present, however, some care has to be taken to avoid such waveforms passing through the meter. The circuit must always be broken at a point where the meter is decoupled on each side. The meter behaves as an inductance, a tuned circuit, or a capacitance, according to the frequency of the waveforms to which it is connected and must therefore not form part of any signal circuit.

The second method is less accurate, because the tolerance of the resistor causes some inaccuracy before measurement starts, and this adds to any inaccuracy in the meter. It does have the very considerable advantage of requiring no broken circuits, however, if the resistor is kept in circuit and lack of accuracy is unimportant if the reading under correct operating conditions is known.

Many elaborate circuits contain inbuilt current monitoring

resistors, decoupled if necessary, so that quick checks of important currents may be made; in some cases several such resistors may be connected through a switch to one point so that a meter may be attached on a set range, the switch rotated and several current checks carried out in a short time.

Complex apparatus, where current checks are frequently useful, require breaking circuits, and where no such provision is made, the service engineer may save much future time by installing a set of monitoring resistors and a switch and noting the readings obtained with the circuit working normally. Preferably, the monitoring resistor should be chosen in value so that the same voltage scale may be used for each. Note that many low voltage, low current transistor circuits *cannot* be monitored in this way because the meter draws more current than can be spared by the circuit.

Measuring Voltage (d.c.)

Most voltages are measured in a straightforward way from a circuit junction to earth, but in a few cases voltages between two other points must be measured. In circuits using p-n-p transistors it is frequently useful to fix the negative side of the meter to the negative line and to use the positive lead to check voltages at the collectors.

Where the voltage to be measured is at low impedance, with currents of several milliamps flowing, there is no more to reading a voltage than clipping on the leads, having adjusted the range switch, and reading the scale. When the voltage is at a high impedance, or if the current which can be delivered is limited in some way, some care must be taken, and the voltage reading may have to be taken by another method. The reason for this is that the meter is a current-operated instrument, and to make it read voltage ranges, resistors are connected in series with the meter movement by means of the range switch.

For example, if the meter requires 1mA for a full-scale reading, and a 10V range is needed, then a resistor is connected in series so that the total voltage across resistor and meter is 10V when 1mA flows. By Ohm's Law, this means a total resistance value of 10kΩ. If the meter movement has itself a resistance of 100Ω, then the extra resistance which has to be connected in series is 9·9kΩ.

It is not difficult to find the full scale current of the meter as this is usually given in the maker's literature or printed on the meter scale. Sometimes, however, it may be quoted as 'ohms-per-volt', from which the actual current may be calculated. For example, 20,000Ω/V means that on the 1V range, the total resistance of the meter circuit is 20kΩ. The current flowing is therefore 1V/20kΩ, which is 50μA.

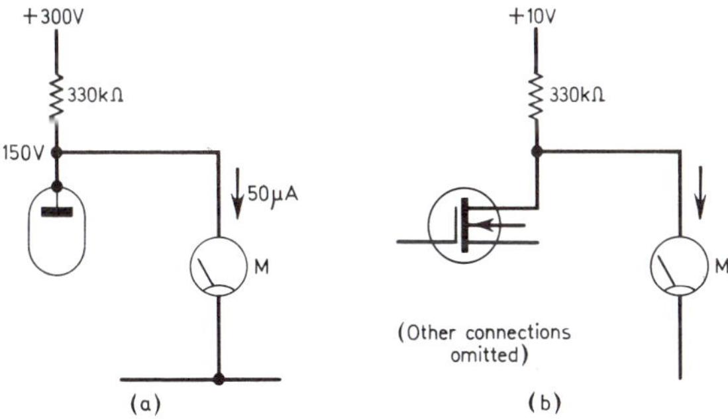

Fig. 1.9 (a) effect of meter on valve circuit, (b) effect of meter on MOSFET circuit.

The effect of a meter on a high impedance circuit can be illustrated by the following example. The voltage at the anode of an audio amplifier valve is to be measured, and the design voltage is 150V, the anode load resistor being 330kΩ (Fig. 1.9a). Suppose, to be pessimistic, that the meter is to be used on the 150V scale and that the meter will then be drawing 50μA. The extra current taken by the meter, assuming that it causes no other effects, will cause the voltage to drop by 330kΩ × 0·05mA, which is 16·5V, so that the meter would read 133·5V. In this case, the error is not too serious, as this amount of difference from the design figure could equally well be caused by a very slight change in grid bias or other conditions.

If this measurement had been taken by a meter using 1mA for full scale deflection, full scale deflection could not have been achieved, since it would have meant a voltage drop of 330V (1mA × 330kΩ) and the actual reading would have been of a low voltage, depending on the potential dividing action of the resistance of the meter and the reaction of the circuit to the lower voltage.

Using a meter of 50μA f.s.d., the measurement was not unreasonable, but if the circuit had been of a MOSFET with a 330kΩ load in the drain and a 10V supply, the use of a 50μA meter would have caused a false reading, since 16·5V could not possibly be dropped across the resistor. Very high impedances are commonly found in the potential-dropping chains of resistors used for instrument c.r.t.'s and photomultipliers, in several types of amplifiers and counters, and in voltmeters designed to measure small voltages from high impedance sources.

Voltages in such circuits cannot be checked with a normal multimeter, and a high impedance instrument (valve or MOSFET operated) is required. The input impedances of such instruments can be of the order of 1000MΩ or more. Very high voltages at high impedance require the use of electrostatic meters.

A.C. Measurements

Most multimeters provide ranges for the measurement of alternating current and voltage. Several important points must be noted about these scales, which depend on rectifying the alternating voltage and measuring the d.c. produced by means of the normal meter action. The sensitivity of the meter is much lower in these ranges, so that high currents are taken, and the scales of the meter are calibrated to read r.m.s. voltages or currents (the values of a.c. which produce the same heating effects as d.c. of the same voltage and amperage figures).

The calibration is for sinewave a.c. only, so that the readings are totally inaccurate if the waveform is anything but sinusoidal. In addition to this, the efficiency of the rectifier starts to drop at frequencies (usually about 10kHz) quoted by the manufacturer, so that measurements cannot be relied on at high frequencies. Moreover, the current transformers used for the a.c. current ranges have a limited frequency response.

Valve and Transistor Voltmeters

The sensitivity of any current-measuring meter may be greatly extended if a d.c. amplifier is used in conjunction with it. This can be used either to provide much greater sensitivity, reading to a few pico-amps in some cases, or it may be used to provide a high impedance at the input, so avoiding the disadvantages of the multimeter in measuring voltages across high-value resistors. The operating procedure for such meters varies, but some features are common.

The amplifier, usually a 'balanced' type (see later) must be set so that no current flows through the meter when there is no input. A control, usually labelled 'Balance' or 'Set Zero' is provided, and adjusted to give zero reading on the meter when the leads are shorted together. When the voltmeter is valve operated, this setting adjustment should not be made until the valves have completely warmed up (allow at least five minutes) and on all amplifier meters, the zero setting should be checked frequently, particularly if there are temperature changes around the meter.

Waves and Pulses

When a voltage varies from one instant of time to the next so that its various voltage levels repeat at definite intervals of time, then we are dealing with a wave or a pulse. The difference between the two waveforms is that the voltage level of a pulse is steady for all but a short interval of time, the period of the pulse, but the voltage level of a wave is continually changing or spending short periods at different values of voltage.

Some of these waves and pulses are generated as a natural result of the laws of electricity; for example, the sinewave is the waveform generated when a coil is rotated in the field of a magnet (the principle of the alternator), and the square pulse is generated when a d.c. supply is switched on and off. Other waves and pulses are not 'found naturally', and must be generated by special methods.

Because the sinewave is mathematically the simplest waveform, and is approximately the one conveniently supplied to us by the domestic electricity authority, it has been customary in the past to think of it as the most important waveform, and to regard it as the source of all the others. Mathematically, this is quite true because any waveform can be constructed by adding a number of sinewaves of related frequencies together, but there seems little reason to labour this point unless we can buy the whole package, maths. and all. From a more straightforward point of view, more can be achieved by taking the squarewave.

The squarewave, which is the wave produced in any switching operation, is shown in Fig. 1.10, together with some of the terms used to specify and describe waveforms and pulses. The first point to note is that there is a voltage 'baseline' round which the squarewave is symmetrical. This is shown as a dotted line which lies midway between the extreme voltages of the wave.

In many cases this baseline is at earth voltage; when it is not, the wave is said to have a z.f. (d.c.) component, equal in the case of the wave shown, to the voltage between the baseline and earth. This is

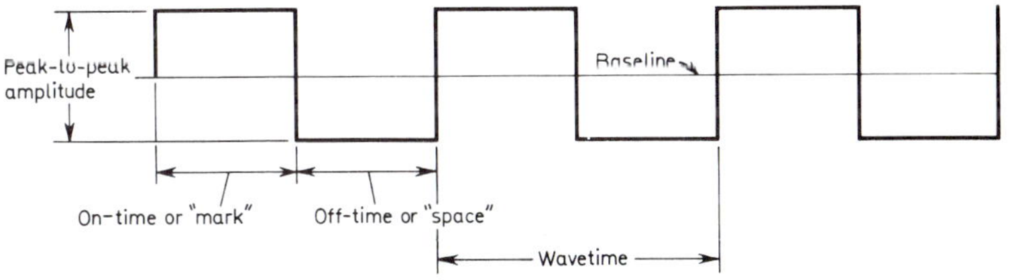

Fig. 1.10 The squarewave.

the voltage value which a moving coil meter, or any multimeter switched to a z.f. (d.c.) range, would record if placed in circuit, provided that the meter did nothing to disturb the circuit conditions.

The amplitude of a waveform is a figure expressed in volts. It may be the peak amplitude, which is the voltage measured from the baseline to one voltage peak, or the peak-to-peak voltage measured from one voltage peak to the other. Such voltage would always be measured with an oscilloscope.

The wavetime is the time in seconds, mS, nS, or μS, whichever may be the case, between one point on the waveform and the next identical point on the next wave. It may be taken from the time when the voltage level crosses the baseline in a positive direction (upwards) to the next positive-going crossing, or between any other easily identified points. Once again, this is an oscilloscope measurement, and most oscilloscopes are equipped to make this measurement fairly accurately, some very accurately with the aid of calibrating waves.

The inverse of the wavetime (1/wavetime), when the wavetime is in seconds, is called the frequency, and the unit is the Hertz, which is the complete wave number per second. For many purposes, the wavetime is a more useful figure, as it is the quantity measured by the oscilloscope, though frequency is measured by frequency meters and digital counters.

In some waveforms, particularly pulses, the width may be important. A squarewave may not be symmetrical, which means that the width of the sections above and below the baseline may not be the same. In an extreme case, such a squarewave would qualify for the name of a train of square pulses.

If we compare one squarewave with another of the same wavetime and amplitude, we may find that the leading edges never coincide,

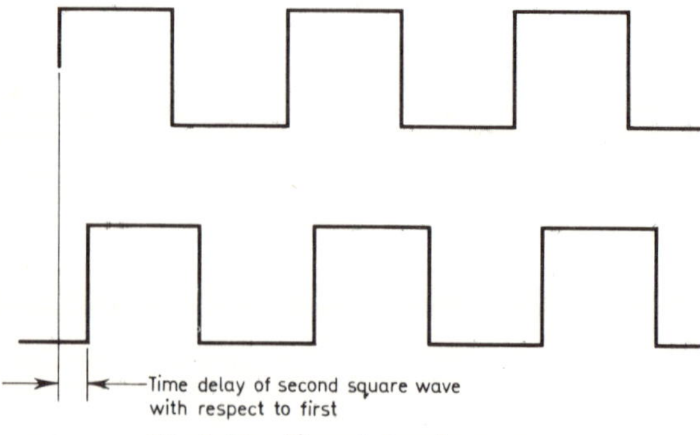

Fig. 1.11 Time delay of a squarewave.

when both waveforms are shown together on a double beam oscilloscope (which shows separately two waveforms in the same time scale). The time difference between the two sets of leading edges is called the *delay time* (Fig. 1.11). When this happens with a sinewave, the difference is usually called *phase shift*, and is stated in terms of degrees rather than in time. Since the oscilloscope measures the shift in time, this may seem rather pointless, but it is related to the way in which a sinewave is formed. Translation is easy:

$$\frac{\text{delay time}}{\text{wave time}} \times 360 = \text{phase shift in degrees.}$$

Phase shift and time delay are not necessarily identical in their effect on a wave. When a squarewave is delayed, its shape is preserved, as is that of a delayed sinewave. There are many circuits, however, which shift the phase of a sinewave, but totally alter the shape of a squarewave. Such circuits shift phase by an amount which depends on the frequency of the signal, so that the effect on a squarewave (which can be imagined as made out of a number of sinewaves of different frequencies) is to change its shape.

Changing Wave Shape

Although a great variety of changes can be carried out on a waveform, there are two of very great importance which can be carried out by passive networks alone. These are termed differentiation and integration, and are of such importance in pulse work that every student should set up an oscilloscope with a squarewave generator and assorted capacitors and resistors to examine these waveforms for himself.

Leaving the methods of producing these effects until later, let us examine the effects of differentiation and integration on a squarewave. The squarewave is made up of vertical and horizontal portions. The effect of differentiation is to exaggerate the vertical portions and practically lose the horizontal portions. The amplitude of the waveform is also greatly reduced during this operation. Integration has the effect of rounding off all the vertical portions, smoothing out the waveform until, in the extreme case, it has become almost z.f. with a slight trace of a.c. 'ripple' noticeable. (See Fig. 1.12).

A sinewave does not change shape when it is passed through these networks, but it changes in amplitude and in phase. Since the shape changes of a squarewave are so much more noticeable when the oscilloscope display is viewed, the squarewave is a much better method of checking the operation of the circuit. Where

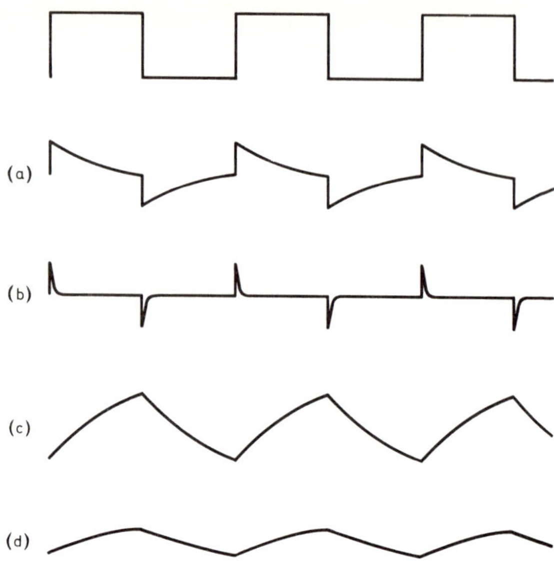

Fig. 1.12 Differentiation and integration of a squarewave. (a) differentiation, (b) severe differentiation, (c) integration, (d) severe integration. Note: the loss of amplitude in severe differentiation or integration would be much greater than has been shown.

instruments are available which can accurately measure the amplitude and phase of a sinewave, the opposite is true.

These forms of diagnosis of what a circuit does are particularly important if an integrated circuit of unknown characteristics is being investigated, as the "circuit" cannot be drawn out by inspection. Fig. 1.13 summarises the effects of differentiation and integration on different waveforms.

Defects of Waveforms

Up till now we have considered perfect squarewaves. As with anything else, perfection is a matter of how closely you look, and a squarewave examined closely with a good oscilloscope may prove to be anything but perfect. Some of the imperfections are due to the fact that the wave must pass through networks (inside the squarewave generator) which differentiate or integrate it; some are due to the imperfect switching action of the active devices used to produce the squarewave. Fig. 1.14 shows in detail some of the imperfections which may exist in a squarewave.

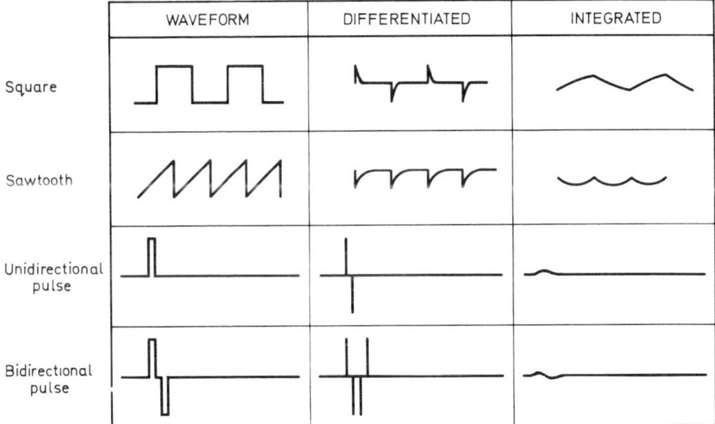

Fig. 1.13 The effect of differentiation and integration on various waveforms.

The rise time measurement detects any integration of the wave. It is the time taken for the voltage to rise from 10% of peak amplitude to 90% of peak amplitude, and is expressed in μS or nS (possibly mS for a very long squarewave). Some engineers use the 5% to 95% amplitude figures for rise time, which makes the rise time seem longer in comparison. The longer the rise time, the more the waveform has been integrated. Overshoot usually is due to inductors in the circuit. The amount of overshoot is measured as the voltage of the overshoot peak above the squarewave peak as a percentage of the squarewave voltage peak.

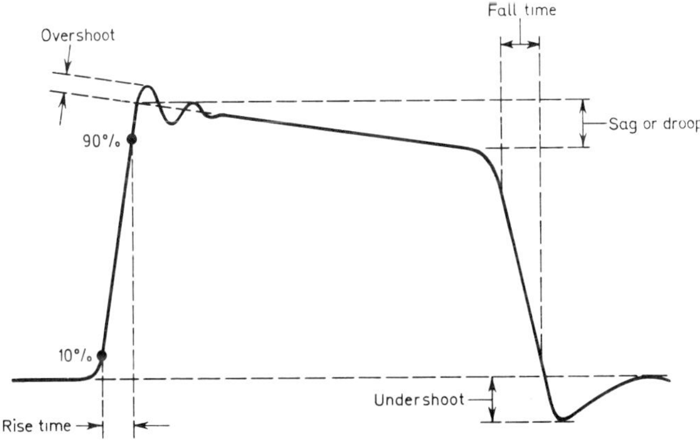

Fig. 1.14 Faults in a squarewave.

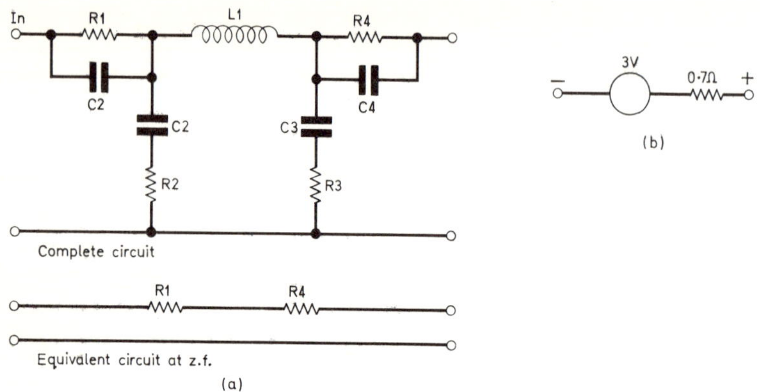

Fig. 1.15 Equivalent circuits. (a) showing how the use of an equivalent can simplify the understanding of a circuit at z.f., (b) equivalent circuit of a battery.

Overshoot time may also be measured. In some cases the overshoot may be the first cycle of an oscillation which persists right along the top of the squarewave. This indicates the presence of a tuned circuit with low damping. Droop or sag is again measured as a percentage of the peak voltage of the squarewave, and is a measure of the differentiation which has taken place.

Fall time is measured between the 10% and 90% levels (possibly 5% and 95%), starting from the sag level and measuring downwards. It also measures integration effects. Finally, undershoot or backswing is measured in the same way as overshoot, is due to the same causes, and may also be followed by a train of oscillations.

Effects of Active Devices

So far we have looked at the effects of what are termed passive linear networks; that is, circuits made up of components which add no power to the signal and whose outputs graphed against the inputs which produce them form a straight line graph. There are passive elements which are non-linear; voltage dependent resistors (the simplest being an ordinary light bulb), voltage dependent capacitors, saturable inductors etc, which cause severe changes to waveforms. The effects of nonlinear passive components cannot be dealt with generally but some special cases are considered later.

Active devices can be used purely to amplify signals, a task so important as almost to merit a book on its own. In addition, active devices can clip, select levels, gate, switch, and assist in integration and differentiation.

Clipping is the removal of one part of a waveform which lies outside a set limit of voltage, current or time. The most common form of clipping is voltage clipping, where the voltage peak of a waveform is removed. This may be in one direction only (unidirectional clipping) or on both positive and negative peaks (bi-directional clipping), and is often carried out by using the fact that an active device amplifies over a limited range of voltage only. Another method is to use a biased diode.

Time clipping is the removal of a portion of waveform at some fixed time after the start of the waveform, and is found mainly in pulse circuits. Level selection is the choice of a waveform above a chosen voltage level, and is a form of gating.

Time gating is the selection of a portion of waveform which occurs between chosen times after the start of the waveform or after a gating signal from another part of the circuit. All of these circuit procedures require the non-linear action of active devices.

Equivalent Circuits

When two different circuits have the same effect on a waveform, they are equivalent as far as that waveform is concerned, and we often find that a complex circuit can be replaced by a much simpler one (in theory) if we only want to find out the effect which it has at one set of terminals.

For example, a battery, which we know consists of a zinc case with a carbon anode, the whole containing the chemicals which cause the battery action, can be thought of electrically as a source of e.m.f. with a resistor (the internal resistance) in series. We do not need to know anything more to design electrically round the battery. As far as the output terminals are concerned, an amplifier is a source of voltage, or current, with a resistor (output resistance) in series; at its input terminals it acts like a resistor only. (Fig. 1.15b).

In many cases it is a great advantage to work with these simplified equivalent circuits. In some cases we can draw equivalent circuits for one frequency or range of frequencies (Fig. 1.15a). A network of capacitors, inductors and resistors behaves at d.c. as if the capacitors were open circuits, the inductors short circuits and only the resistors actually present. For very short rise time waveforms, the capacitors seem to be short circuits, the inductors large impedances and the resistors small by comparison. Some networks can be better understood if drawn as an equivalent circuit, and their effect on the circuit around them can certainly be better understood.

It should be clear, however, that the equivalent circuit does not describe how a circuit works, merely how it behaves, and in some cases its use may be misleading if we try to use it where it was not intended, such as when the low-frequency equivalent of a transistor is used to predict its behaviour as a switch.

CHAPTER TWO

RESISTORS

ANY SUBSTANCE WHICH OBSTRUCTS the flow of electric current equally at any frequency, including zero, and which releases heat according to the formula $W = V \times I$ (where W is the power in watts released as heat) can act as a resistor. In the early days of radio, amateurs made resistors by drawing pencil lines on the wooden boards on which circuits were constructed; a bolt, washer and nut at each end of the line served as the leadouts of this resistor. Low values of resistance were made by winding spirals of thin wire round ceramic formers.

The use of carbon (the pencil line) and thin wire as materials survives in slightly different forms, but several new materials for resistors and many new methods of using the old materials have come into use recently.

Resistor Characteristics

The two most obvious characteristics of any resistor are its resistance in ohms and its maximum dissipation in watts, but only the first is marked on most resistors. The Ohm is defined as the value of the resistance which has one volt between its ends when a current of one ampere flows; the equation connecting the three quantities is, of course, Ohm's Law, and the dissipation equation (Joule's Law) is quoted in the opening paragraph.

The resistance value marked (see colour code, Fig. 2.1) is not, however, the exact value of resistance which the user of the resistor might measure. In most applications we need resistors which are only approximately of the indicated value, and so each resistor has a 'tolerance' figure which indicates how far from its marked value the actual resistance may be.

In many cases, this tolerance figure is colour coded or marked on the resistor, popular values being 20%, 10%, 5% and 1%. Sometimes the tolerance is even closer, such as in the case of resistors for meter shunts. In addition to the tolerance limits, resistors made by some processes may change in value during storage, during use and with changes in temperature.

Preferred Values

When resistors are manufactured, certain preferred values are aimed at. These preferred values are based on a scale starting at 1 and with each step in geometrical progression, meaning that each number of the scale is a constant factor times the last number. In practice, this works out to give a scale where each number is roughly the same percentage above the last one.

For example, in the 20% series, each number is $\sqrt[6]{10}$ above the previous one (the sixth root of ten). The value (approximate) of $\sqrt[6]{10}$ is 1·46, and the numbers are rounded off to give the familiar series, 1, 1·5, 2·2, 3·3 etc., each of which is about 40% above the previous one. The 10% series uses as its multiplier $\sqrt[12]{10}$ (about 1·21) and gives a 20% step up; the 5% series uses the factor $\sqrt[24]{10}$ (1·10) and gives a 10% step up.

The effect in each series is that, if we aim to manufacture resistors of preferred values given by the numbers of the series, then *any* resistor which we make must fit in the series. If we aim at making 2·2kΩ and find that we are getting 1·9kΩ, then we can use the 1·9kΩ as 2·2kΩ in the 20% series, or as 1·8kΩ in the 10% series. The preferred values in the 20%, 10% and 5% series are shown in Table 2.1.

Higher values are simply multiples of the figures shown; for example, the next value above 10Ω in the 20% series is 15Ω, the value above 100Ω is 150Ω etc. The same number series is used for capacitors. Note that it is pointless to search through a box of 20% tolerance resistors, measuring each, in an effort to find one within 5% of a preferred value. The 10%, 5% have already been sorted out by the manufacturers, and the 20% tolerances will be just that.

Changes in Value

Changes in resistor value affect both the designer and the maintenance engineer. The designer must decide if the circuit is likely to be affected by a large tolerance or by a change in value, and the maintenance engineer must decide, on measuring the value of a resistor in a circuit, whether this lies within the tolerance allowed or has changed to a serious extent.

Consider for example, the situation of Fig. 2.2. A resistor is used here to set the screen voltage of a pentode valve. The design is based on a line voltage of 300V, a screen voltage of 200V and a

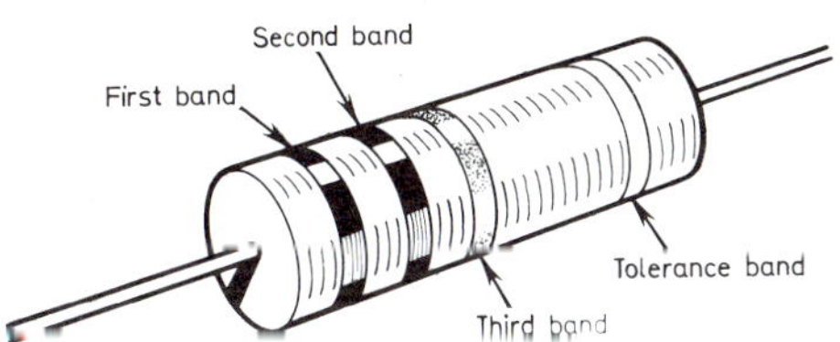

The first band gives the value of the first significant figure, the second band gives the second significant figure, the third band gives the number of ciphers (zeros). Example: brown, black, red = 1–0–00(2) = 1,000; or brown, green, yellow = 1–5–0000(4) = 150,000. A salmon pink body indicates a high stability resistor.

Figure	**Colour**
0	black
1	brown
2	red
3	orange
4	yellow
5	green
6	blue
7	violet
8	grey
9	white
5%	gold
10%	silver

(tolerance band colours only)

Fig. 2.1 Resistor colour code.

screen current of 10mA. The voltage to be dropped across the resistor is 100V, so that, using Ohm's Law with units of mA and kΩ, we have 100 = R × 10, giving a value for R of 10kΩ. The cheapest available resistors are 20% tolerance; 20% of 10kΩ is 2kΩ, and the actual resistor value may be 8kΩ or 12kΩ.

How does this affect the design voltage? If the line voltage remained at 300V and the current remained at 10mA, the voltage drop for these extremes of tolerance would be 80V or 120V, making the screen voltage 220V or 180V. This, however, does not happen, for the screen current of a valve is affected by the voltage, and the current would not remain at 10mA. It would, in fact, increase at the higher voltage, causing a greater voltage drop than the 80V which we calculated, and would be less than 10mA at the lower voltage, causing a smaller voltage drop. The current would change, in fact, so as to counteract the effect of the resistor tolerance.

The designer would conclude that a 20% tolerance would have little

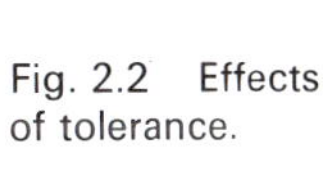
Fig. 2.2 Effects of tolerance.

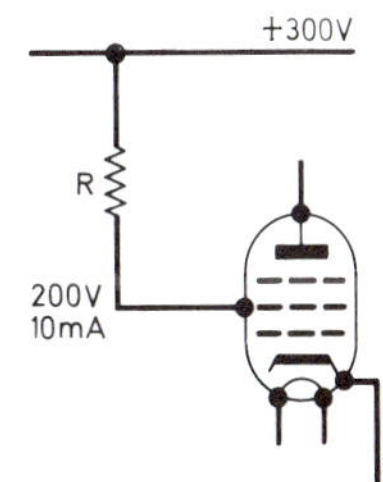

effect unless he particularly wanted to hold the *current* steady; he would also conclude that changes of resistance during use would have little effect. The service engineer, finding that the voltage was not exactly 200V would not consider this to be a cause of trouble unless the shift were much larger, say to 280V or to 120V.

Maximum Dissipation

The maximum dissipation of a resistor is the amount of electrical power which it can continually change to heat without damage. When a resistor dissipates power, it becomes hotter than its surroundings, and this causes a change in its resistance value. If the dissipation is too great, or if there is no way of allowing heat to escape (in the case of a resistor closely surrounded by other components), the resistor may change considerably in value, and may not recover to its marked value when it cools. In extreme cases the high temperature may burn out the resistor, so causing total failure.

In many cases, a service engineer would notice blackening or blistering around a resistor running hot, but in some cases where a very large, short duration, overload had caused failure there might be no external signs of damage. The warmer the air round the resistor, the less chance it has of getting rid of the heat which it dissipates and the more likely is the chance of overheating.

The maximum dissipation of resistors is therefore quoted at a maximum temperature, usually 70°C (158°F) above which the power dissipated in the resistor must be reduced. 70°C is a high air temperature (water boils at 100°C) but such temperatures can be achieved in the air inside cabinets, especially where valves are used in the circuit, or where several high-dissipation resistors are bunched together. The designer has to calculate the dissipation and place the resistor so that the air temperature round it is not too high.

As an example of dissipation, consider the screen resistor of the last example, which will be working in a position of high temperature near a valve. It is designed to drop 100V at 10mA, and its dissipation will be 1W (volts × milliamps gives dissipation in mW, and 1000mW is 1W). At this point a decision has to be made. One watt of dissipation could be handled by a one watt resistor, since some makes are generously rated, or it might be safer to use a resistor rated at $1\frac{1}{2}$W or 2W.

A cautious designer, remembering that resistance changes and dissipation changes with it, would specify the higher dissipation resistor regardless of cost. This approach would be used in any high cost electronic equipment. In domestic radio and television, however,

20%	10%	5%	20%	10%	5%
1·0	1·0	1·0	3·3	3·3	3·3
		1·1			3·6
	1·2	1·2		3·9	3·9
		1·3			4·3
1·5	1·5	1·5	4·7	4·7	4·7
		1·6			5·1
	1·8	1·8		5·6	5·6
		2·0			6·2
2·2	2·2	2·2	6·8	6·8	6·8
		2·4			7·5
	2·7	2·7		8·2	8·2
		3·0			9·1

Table 2.1—TOLERANCE SERIES, showing standard values in each series.

such an approach could price a receiver too high, and minimum dissipations are used.

This approach is justified, for a resistance change in a TV set may only cause a slight change in, say, field height, which can be corrected by a preset control; but a similar change in resistance in a computer circuit might cause a series of errors costing thousands of pounds.

As far as the serviceman is concerned, he must decide whether the failure of a resistor is due to a fault somewhere else causing excessive current through the resistor (for example, a failure in the bias circuit of a transistor can cause excessive current in the collector circuit) or simply due to the resistor being marginally rated.

No designer is perfect, and if there is no obvious reason for resistor failure where ratings are close, the service engineer should always replace with a component of higher rating. It is worth noting that American domestic equipment often uses resistors working at their limits, and it is often worth while to replace all the over-rated components.

Carbon Composition Resistors

These are the most common resistor type encountered. They are traditionally made by pressing a mixture of carbon, ceramic dust, and resin, at high temperature, into the form of a rod which is then fitted with metal caps holding the end wires. This rod may be left as it is, giving an uninsulated resistor or, more commonly, plastics coated in the case of an insulated resistor.

When a particular value of resistor is aimed at, the end product is a large number of resistors whose value varies round the aiming value and those are then measured and sorted. The greatest number are between 10% and 20% out; some are between 5% and 10% and so on. It is in this way that resistors of various tolerances appear; they are not manufactured to a particular tolerance, and it is wrong to suppose that resistors of, say, 1% tolerance can be extracted from a box of 20% resistors by measuring each one. The manufacturers have already done the sorting operation and the price for each tolerance reflects the comparative numbers which can be sorted out.

Composition resistors in general have values ranging from 2·2Ω to 22MΩ, operate at temperatures between −40°C (−40°F) and + 70°C (158°F) with no change in dissipation rating, can withstand 1500V between core and outside in the case of insulated types and have power dissipation rating from 1/25 watt up to 3W depending on type. Maximum change on soldering should be 2% maximum change with high humidity, 15½%, maximum shelf-life change 5%.

Noise levels are quoted as μV of noise per volt of signal or d.c. applied to the resistor and are approximately 2μV per volt for values below 1kΩ, 3μV/V for 10kΩ; 4μV/V for 100kΩ; 5μV/V for 1MΩ, etc. (going up by steps of 1μV for each tenfold increase in resistance).

Wirewound Resistors

The resistance of a piece of wire increases with increasing length and rising gauge number (the higher the gauge, the less the diameter). Different metals have different resistivities; that is a set of bars of different metals, all bars being of exactly the same dimensions, would have very different resistances, silver having least resistance of the metals and an alloy of nickel and chromium about the highest.

Even if an alloy of very high resistivity is used, however, a considerable length of wire is needed. One foot of nickel-chromium alloy one thousandth of an inch thick has a resistance of 900Ω, and we shall therefore need 10ft for 9kΩ and 100ft for 90kΩ. For this reason, the wire must be wound round some sort of former, and if we can wrap a lot of wire round this former we can use thicker wire and increase reliability.

This is not entirely straightforward because a coil of wire acts as an inductor as well as a resistor and its impedance increases as the frequency of the current increases. To avoid this, the wire is wound in opposite directions by doubling the length of wire over and winding the double layer on.

Wirewound resistors are used for very low resistances and for resistances which must dissipate more power than can be handled by the carbon composition types, dissipations of several hundred watts being quite common in industrial electronic equipment. They should be avoided for high frequency signal paths, or bypassed by a capacitor if this is possible.

Wirewound resistors are usually coated with a ceramic insulating material, and the value and power dissipation are written or printed on this. Colour coding is ineffective as the colours change with the heat; the surface temperature is 250°–300°C (app. 500°–600°F) at full dissipation.

Film Resistors

This is a much more modern type of resistor, used to a very large extent on electronic equipment of high reliability. The thinner a sheet of material can be made, the higher the resistance a strip of given dimensions has, and if we make the material in the form of a film so thin that it is transparent then the resistance can be very high.

Such a thin film is not self-supporting and so it must be formed on some surface—a rod of porcelain or glass is very suitable. The value can then be precisely adjusted to a higher resistance by cutting a spiral track in the film; the final value can be about two hundred times the resistance of the film as first made. The films may be of metal, of carbon, or of tin oxide, each having different characteristics as regards change of resistance with temperature, voltage, etc.

Film resistors have very close tolerances, and change resistance only very slightly with change of temperature and other conditions. They are produced in tolerances ranging from 0·1% upwards (compare the carbon 'close tolerance' value of 5%). For the most critical specifications of aircraft navigational systems, metal film resistors are used. In other applications film resistors of other types have replaced the older H.S. (High Stability) rating which used to be used for carbon composition resistors run at low dissipation and made to high engineering standards.

Temperature Coefficients

The resistance of a sample of any material will change with changing temperature. If resistance increases as temperature increases, as is the case with most metals, and decreases as temperature is decreased, then the material is said to have a positive temperature coefficient (PTC); if the resistance decreases as the temperature rises, then the temperature coefficient is negative (NTC).

The figure quoted for temperature coefficient enables the change of resistance to be calculated. For example, if a resistor is quoted as having a PTC value of 200ppm/°C, this means that its resistance changes by 200 millionths (ppm meaning parts-per-million) of its nominal value for each degree Celsius change of temperature. Suppose that the nominal value is 100kΩ, then a 50°C rise in temperature causes a resistance change of 200 × 100 × 50/1,000,000kΩ, which is 1kΩ. Another way of calculating this is that 200ppm represents a 0·02% change per degree. Over 50°, this means a 1% change, and 1% of 100kΩ is 1kΩ. As the temperature coefficient is positive, the resistance increases by 1kΩ for the 50°C rise in temperature.

A designer may use PTC or NTC resistors to ensure that a circuit remains stable under conditions of changing temperature. For this reason, resistors must not be changed indiscriminately, especially in oscillator circuits, except to check whether a change is needed. The resistor finally used to replace one which is marked with a definite type-number must be another of the same type-number. This does not apply to ordinary colour-coded resistors.

Some materials have very large temperature coefficients and are used deliberately for temperature measurement, temperature compensation, or time delay. One type of such material consists of a mixture of metal oxides, and the component is known as a thermistor. The resistance of a thermistor decreases very rapidly (for example from 3kΩ down to 50Ω) as temperature rises.

High Frequency Effects

We have mentioned that it is difficult to ensure that a wirewound resistor does not behave as an inductor. Resistors of all types also have stray capacitance, and their impedance becomes less at high frequencies in a way which depends on their shape and size rather than on their resistance. As a rule of thumb, carbon composition resistors of 220Ω start to reduce in impedance above 1MHz, 47kΩ at about 5MHz, 4·7kΩ at 60MHz.

Voltage

It is bad practice to drop more than 500V across a resistor except in the case of the specially made long resistors designed for high-voltage power supplies. Even if the power dissipation is not exceeded, there is a danger of flashover at higher voltages, change of resistance due to voltage becomes more serious, and atmospheric conditions cause leakage along the small surface of the resistor. Where a

Prefix	Abbreviation	Value	Index
Giga-	G	1,000,000,000	10^{9}
Mega-	M	1,000,000	10^{6}
Kilo	k	1,000	10^{3}
Milli-	m	1/1,000	10^{-3}
Micro-	μ	1/1,000,000	10^{-6}
Nano-	n	1/1,000,000,000	10^{-9}
Pico-	p	1/1,000,000,000,000	10^{-12}

Prefixes are used to avoid the printing of very large or very small numbers on component lists or on circuit diagrams. Thus, 4,700Ω is printed as 4·7kΩ; the Ω sign is often omitted, so that the value is shown as 4·7k.

Because of the extensive use on the Continent of a comma as a decimal point, commas are often omitted when numbers such as 2500 are printed, a space being left instead. Also becoming more common is the use of the Ω sign or one of the multiple abbreviations, instead of the decimal point as in 1Ω5, 2k7, 3M3 (1·5Ω, 2·7kΩ, 3·3MΩ) which avoids any confusion between previous uses of points and commas.

Table 2.2—PREFIXES AND ABBREVIATIONS

resistance must connect two points with a large voltage difference, it is preferable to connect a large number of resistors in series with less than 500V across each.

Noise in Resistors

In all substances, even at the very low temperature of liquid helium (−270°C), the electrons and other parts of the atom are vibrating, and the extent of the vibration increases with increasing temperature. Movement of the electrons is the equivalent of an electrical signal, and the vibration, which takes place randomly and not with any fixed frequency, is what we term electrical noise.

Noise in active devices, where electrons are arriving randomly at electrodes, is usually more important than noise in passive devices, but carbon composition resistors can have high noise signals due to their construction. For this reason, the first stage of a high gain amplifier may specify low-noise resistors, so that the noise which is amplified by the remaining stages is as low as possible.

Wirewound and film resistors have lower values of noise. The actual noise voltage generated depends on the resistance value and the temperature of the resistor and rises with rising resistance and temperature, and is greater for wide bandwidths than for narrow bandwidths.

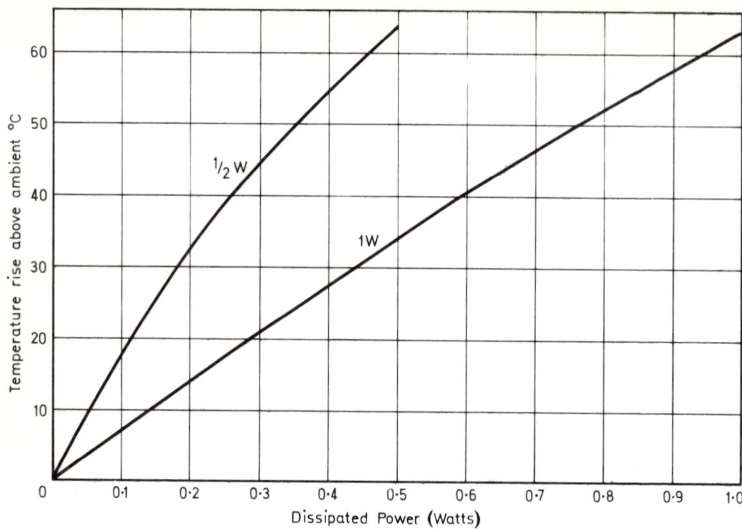

Fig. 2.3 Graph of temperature rise against dissipation. (*courtesy Morganite Resistors*).

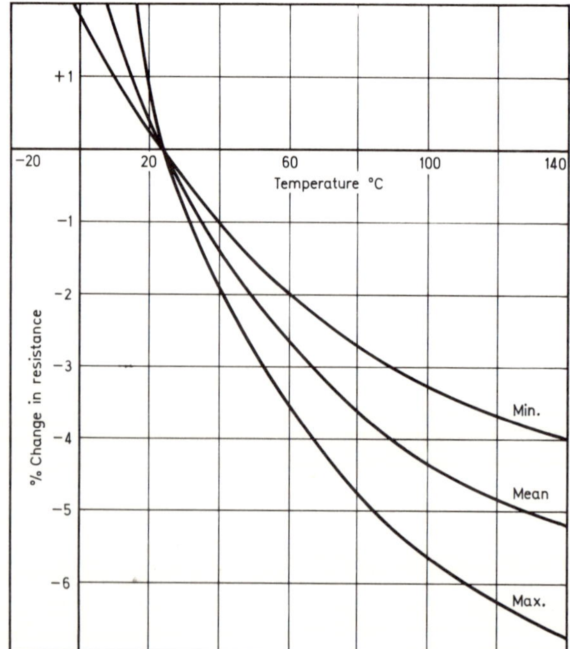

Fig. 2.4 Effect of temperature on a 100kΩ resistor. (*courtesy Morganite Resistors*).

Other Points on Fixed Resistors

It is often useful to know approximately what temperature rise is caused by the dissipation in a resistor. The table of Fig. 2.3 shows the temperature rises produced by various dissipations in $\frac{1}{2}$W and 1W resistors produced by one manufacturer. Note that the figure quoted is temperature *rise*; if the temperature round the resistor is 60°C before there is any dissipation in the resistor, then the temperature change produced by the resistor is added to this figure to give the temperature which would be measured on a thermometer placed next to the resistor in operation. It can be seen from the table that a resistor working at full dissipation can produce an extra 60°C, a total of 120°C which would require derating of the components next to the resistor.

Fig. 2.4 shows the effect of temperature on one example of carbon resistor. Since the value chosen is 100kΩ, the percentage figures of change can be read directly as kΩ. These changes are fairly typical of carbon composition resistors. Fig. 2.5 shows the characteristics of some thermistors, indicating how the resistance varies with extremes of temperature and power dissipated.

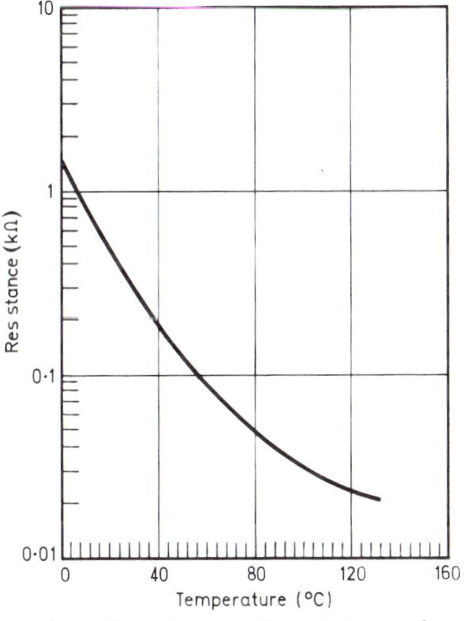

Fig. 2.5 Resistance of a thermistor at various temperatures. (*courtesy RS Components*).

Another type of resistor, the non-linear resistor (NLR) has a resistance which depends to a considerable extent on the voltage across it. For example, an NLR might have a resistance of 10kΩ at 10V and 1kΩ at 100V. This type of resistor can be used as a voltage stabiliser because it passes higher currents as the voltage across it becomes higher. Such resistors are found under the trade name of *Metrosil*.

Lamp bulbs are sometimes used as non-linear resistances; the resistance of a glowing lamp filament can be ten times that of a cold filament. Note that this is because of the PTC rather than variation of resistance with voltage.

Variable Resistors

If a resistor, made by any process, is provided with a third contact which rubs over the resistive material and makes contact, the resistance between the movable contact and either of the fixed contacts is variable. Such an arrangement can be described as a trimmer, a variable resistor, or as a potentiometer according to its applications.

It is important to be able to distinguish the names. The term variable resistor is used when one fixed contact and the moving contact are wired in circuit, the other fixed contact not used, and the control for adjustment of resistance is accessible for day-to-day adjustment. A potentiometer uses all three (sometimes other fixed contacts are provided) contacts in circuit so that the resistor acts as a potential divider (see later.). Again, the control is accessible.

A trimmer may be used either as a potentiometer or as a variable resistor, but is set in value during the test period of a piece of equipment and seldom, if ever, altered afterwards. Trimmers are often used to set the working range of variable resistors or potentiometers, and are often referred to as preset controls.

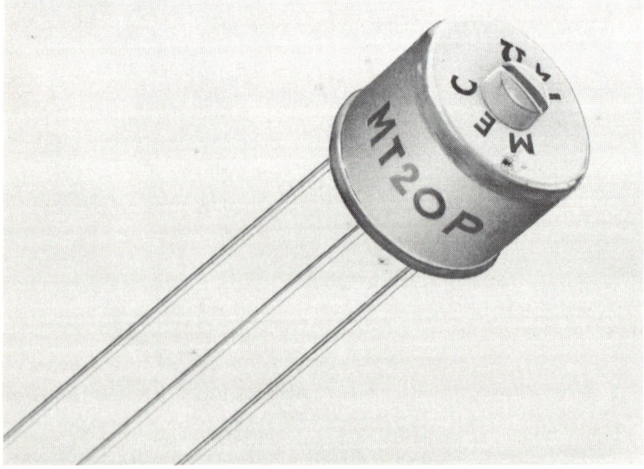

Fig. 2.6 Miniature wirewound trimmer, diameter 8.8mm (*Miniature Electronic Components Ltd.*)

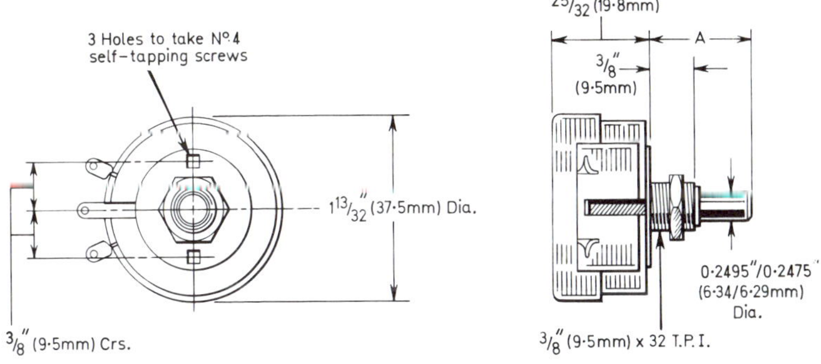

Fig. 2.7 Standard wirewound potentiometer. (*Colvern Ltd.*)

Trimmers are often made simply as carbon composition bars with a third contact. This method of construction keeps costs low and ensures good ventilation. More often, though, the composition element is made circular, with a fixed contact at each side of a gap in the circle. The moving contact is a spring leaf which is attached to a shaft passing through the centre of the element (Fig. 2.6).

Potentiometers are made mainly in the circular form, although flat, slider-type construction (like the faders on professional mixing equipment, for example) is becoming popular in certain areas of domestic audio equipment. Most use elements of carbon composition or wire, the wire being wound round a flexible insulating rectangular former which is then bent into circular form (Fig. 2.7), or round a ceramic former in the case of potentiometers intended for high dissipation. Until recently, wirewound potentiometers were essential for dissipation of greater than $1\frac{1}{2}$W, but new elements based on materials called *Cermets* (ceramic metals) are now being used.

Helical potentiometers use elements shaped like the coils of a spring (Fig. 2.8). The moving contact not only turns round but moves down as if on a screw thread (which it is), and the control knob can be moved through several revolutions, ten turns being fairly common. The advantage of this contruction is that a very fine variation of resistance can be achieved without the need to gear the shaft of the potentiometer. The addition of a digital dial to such a potentiometer is the basis of many measuring instruments.

Peculiarities of Potentiometers

Potentiometers (as with variables and trimmers) must be specified by nominal resistance and power dissipation, just as fixed resistors are specified, though the preferred range of values is not used for potentiometers or other variables. In addition, the designer must specify law, and in some cases resolution. The law of a potentiometer describes the way in which the resistance between one fixed contact and one moving contact varies as the control shaft is turned. If equal amounts of turning (10°, say) give equal steps of resistance all the way along the sweep of the track, then the law is said to be linear, because the graph of resistance against shaft angle is a straight line.

This form of law is not always convenient, and there are such laws as log, antilog, semilog, B-law, and such specialised types as sine and cosine, each of which has special characteristics for a particular purpose. For example, the human ear has a logarithmic response, meaning that the loudness heard increases as the logarithm of the amplitude of the noise.

A potentiometer used as a volume control on an amplifier must therefore increase gain at a rate which will compensate for this and make equal amounts of angle seem to cause equal steps in volume; this requires a log-law potentiometer. When a potentiometer is replaced, it is therefore important that it should be replaced by another of similar law.

The resolution of a potentiometer is unimportant in most radio and TV applications, but is very important in the precision potentiometers used in instruments. The resolution is the smallest step of resistance which can be achieved. In a wire-wound potentiometer it is the change of resistance when the moving contact travels on from one turn of wire to the next. The contact may cover several turns, but still can be moved by the amount of one turn. In a cermet potentiometer, the resolution is limited only by the minimum movement of the shaft which can be achieved.

Carbon composition potentiometers are not used where high accuracy and high resolution are required. Composition potentiometers for domestic entertainments equipment are often rated at temperatures of 20°C to 40°C, and not at the 70°C usually quoted for industrial use.

Tolerances of composition potentiometers are nearly always 20%, maximum working voltages range from 50V (d.c.) for miniature edge mounted presets to 500V for moulded track types. For alternating voltages, roughly 2/3 of these values should be taken. Maximum noise is 47mV, much higher than for fixed resistors; presets may have noise levels up to 100mV.

Minimum resistances (at one end of track) can be as high as 50Ω for moulded track types, and 0·5% of resistance (5kΩ for a 1MΩ) for other potentiometers. Presets can have high minimum resistances (3%) since minimum resistance is of little importance in presets.

It is important to remember these minimum resistances when measuring potentiometers on a resistance meter.

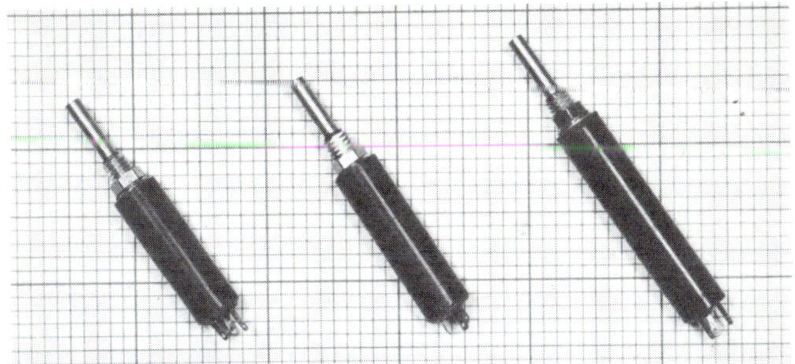

Fig. 2.8 Miniature helical potentiometers, with three, five and ten revolutions. (*Miniature Electronic Components Ltd.*)

Limitations

Very fine wire and very thin carbon tracks cannot be used, since the moving contact would soon wear them out. Wirewounds are made in the range of values 2Ω to 150kΩ, and carbon composition in the range 100Ω to 5MΩ. The range of cermet tracks is wider than that of wirewound, but not so extensive as that of composition.

Wirewound potentiometers are usually inductive, and cannot be used for signals above about 10kHz, although low-inductance types are available for rather higher frequencies. It is not therefore justifiable to replace a burned out carbon potentiometer in a signal circuit by a wire-wound. Carbon potentiometers used as volume controls or gain controls should not have z.f. current flowing through them, since this causes contact sparking, intermittent contact and noisy operation. Grease and dirt are the main enemies of potentiometer tracks, particularly of open tracks, and regular cleaning is advisable. A variety of cleaning fluids is available, but it is wise to check the effect of the fluid on other components before use.

Other Trimmers

Flat, or edge, trimmers are being increasingly used. The resistive element is a flat slab coated with cermet, or wirewound, and the contact is moved along by a lead-screw. A spring clutch is provided to prevent damage due to overscrewing. Such trimmers have high resolution and can be stacked closely together.

Subminiature trimmers can be obtained in a variety of styles for any form of mounting. One popular form is similar to a transistor in size and in casing, and can be mounted directly on a printed board.

Measurement of Resistance Value

For many purposes, because of tolerances, only a fairly approximate value of resistance is needed and an ohmmeter can be used. An ohmmeter consists of a battery, a variable resistor (whose control is labelled 'zero set'), a meter and a pair of terminals. With the terminals shorted, the variable resistor is set to give full scale deflection on the meter.

With the terminals open-circuit, the meter current is zero and this end of the scale is labelled 'infinity'. Any resistor which is now connected to the terminals will cause some value of current to flow, the amount depending on the resistance in circuit, and this current will be indicated on the meter, which can now be rescaled in terms of resistance.

The advantages of an ohmmeter are that it is quick and simple to use. The scale is, however, the 'wrong way round', and is very cramped at the high resistance end. A very sensitive current meter is needed if resistors of 1MΩ or over are to be reliably measured, and very low resistance values present a problem of excessive battery drain. Usually, several scales have to be used, requiring individual setting of zero.

More precise measurements of resistance are carried out with a measuring bridge. Bridge circuits are dealt with in detail in the companion volume to this book (*Understanding Electronic Circuits*); the principle is that of comparing one value of resistor with another. This may be done using a.c. or d.c., and does not require a calibrated meter. A measuring bridge is as accurate as its built-in resistors permit. As these few resistors may be wirewound and trimmed by hand to the final value required, this accuracy may be very good indeed.

For most purposes, resistors mounted in a circuit have to be measured with at least one end detached from the circuit. Some bridges exist with a built-in 'back-off' so that a resistance value can be measured with the resistor in circuit.

CHAPTER THREE

CAPACITORS

WE NOTED IN CHAPTER 1 that the units of negative electric charge, the electrons, repelled other electrons and attracted positively charged particles. A capacitor is a device which uses this principle to store electric charge. Imagine two parallel metal plates separated by air. If we apply a voltage between the plates from a battery, we have ensured that one plate has a surplus of electrons. The other plate must then have a surplus of positive charge, because the electrons are repelled from it by the negative charge on the first plate. (Fig. 3.1a).

When the battery connections are removed, we end up with the static situation of excess electrons on one plate, too few on the other. There is still a voltage, equal to the battery voltage, between the plates, because the excess and deficit of electrons causes one plate to be positive and the other negative, and if we now connect the two plates by a wire, a current of electrons will flow until the number of electrons on each side is balanced.

The total amount of charge which flows when the plates are connected is calculated by multiplying the current which flowed by the time for which it flowed, not so straightforward a task as might be thought because the current is not a constant value. We find that,

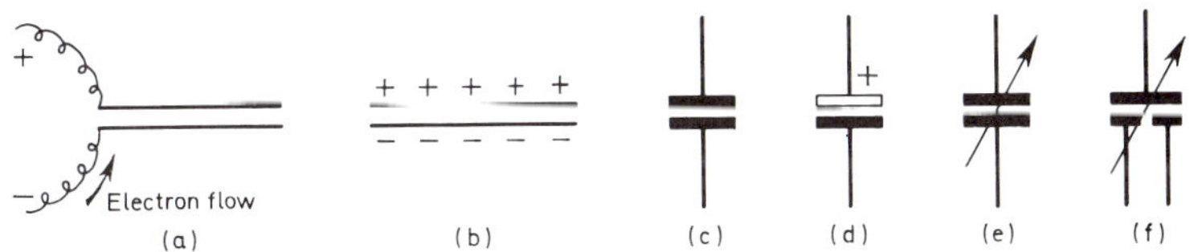

Fig. 3.1 (a) parallel plates being charged by wires from battery, (b) plates charged, (c) symbol for capacitor, non-electrolytic and fixed, (d) symbol for electrolytic capacitor—note polarity, (e) symbol for two-vane variable capacitor, (f) symbol for three-vane variable capacitor.

no matter what voltage we used, if we calculate the figure Charge on plates/Voltage between plates, then the figure is always the same for the same two plates at the separation distance which we set. This Charge/Voltage figure is called the capacitance of the plates, and the arrangement of plates is called a capacitor.

When electrical units were being devised, the unit of capacitance was, and still is, the coulomb (the charge which flows when a ampere of current flows for one second), and the unit of voltage was, and is, the volt. When coulombs are divided by volts to give the unit of capacitance, called the farad, this unit turned out to be much too large for practical purposes.

In electronics we use the microfarad (μF), which is one millionth of a farad (10^{-6}F), the nanofarad (nF), which is one thousandth millionth (10^{-9}F) of a farad, and the picofarad (pF), which is one millionth of a millionth of a farad (10^{-12}F).

Behaviour of Capacitors

A d.c. voltage can charge up a capacitor. If a battery is connected to a capacitor, current flows from the battery until the capacitor is charged, but though current has flowed through the wires into the capacitor none of this current has flowed through the capacitor, as might be expected as the plates are insulated from each other. The current can only flow for a very short time, until the voltage of the plates equals the voltage of the battery.

The total amount of charge which has flowed to charge up a capacitor from 'empty' to 'full' depends on the capacitance and the voltage to which it is charged, the amount in coulombs is given by: Charge = Capacitance × Volts, noting that capacitance must be stated in farads if the charge is to be in units of coulombs.

This amount of charge will be given out again if the capacitor is short-circuited, and can be quite considerable for large, high voltage capacitors. It is a wise precaution to clip a piece of wire between the terminals of high capacitance, high voltage, capacitors before carrying out any work near them.

In the case of d.c., the capacitor charges up and that is the end of the story. If we continually reversed the voltage, however, the amount of charge, given by $Q = C \times V$, would be flowing into and out of the capacitor at each reversal. As far as we are concerned, this is a current, an alternating current.

When an alternating voltage is applied to a capacitor, an alternating current flows, and the amount of the current depends on the size of the voltage and the size of the capacitor. In this way, the capacitor is behaving like a resistor, and even obeys a form of Ohm's

Law, but in three other ways the effect is quite different. For one thing, the higher the frequency of the voltage the less time the charge has to move.

Since charge is Current × Time, less time means more current for the same charge, so more current flows. Formally, this is expressed by saying that the reactance of a capacitor (the factor which corresponds to resistance when we write $V = X \times I$ for a capacitor, where X is the reactance.) decreases as frequency increases.

The second difference is that no power is dissipated when a.c. flows 'through' a capacitor, and the reason for this will be more apparent when we look at the third difference.

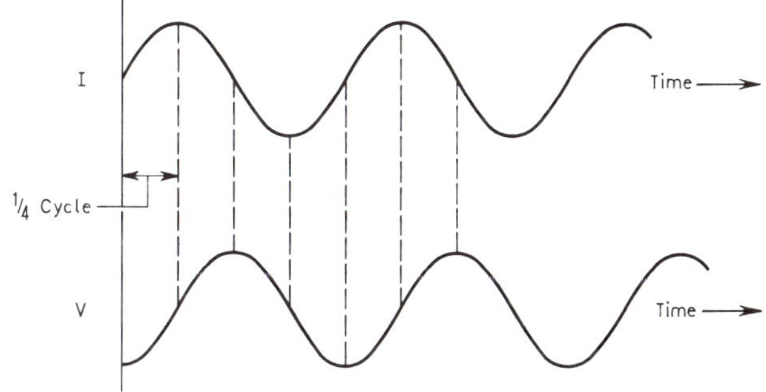

Fig. 3.2 The voltage across a capacitor is $\frac{1}{4}$-cycle later in reaching its peak than is the current, and the voltage is $\frac{1}{4}$-cycle behind current at all times.

The voltage across the capacitor is at maximum when the capacitor is fully charged, which is when the current has (momentarily) stopped flowing. Maximum current and maximum voltage never happen at the same time, and we find that maximum current is always $\frac{1}{4}$-cycle (90°) before maximum voltage (see Fig. 3.2). It is because maximum current is at zero volts and maximum volts at zero current that there is no power dissipated in a perfect capacitor, because the product of volts and amps is always zero when we average them over one cycle.

Construction of Capacitors

For a capacitor made of flat plates, the capacitance increases as the area of the plates increases and as they are brought closer together. The capacitance is also increased if some material, other than air, which does not conduct electricity, is used to space the plates. The amount by which capacitance is increased over an airspaced capacitance in this way is a measure of the permittivity of the substance. The whole equation for capacitance is: $C = 28 \times k \times A/(\pi \times t)$, or about $9kA/t$, where C is in units of pF, k is the permittivity (between 1 and 20), A is the area of each (equal) plate in square metres, and t is the spacing between the plates, also in metres (Fig. 3.3).

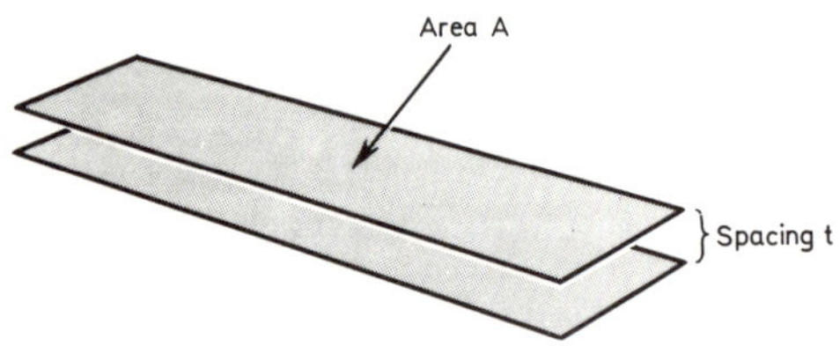

Fig. 3.3 Calculating the capacitance of a pair of plates.

Very small capacitance values are made simply as parallel plates or wires, with materials (dielectrics) of high permittivity between them, but high capacitance values would be inconvenient to make in this way, as they would be too large. A sandwich of plate, in the form of thin aluminium foil, dielectric (thin waxed paper or plastics sheet) and another aluminium foil is wrapped round and round to form a cylinder from which a lead is taken to each foil, and the whole is encased in plastics. (Fig. 3.4).

In the wrapping process, the outer foil would short against the inner, so a second layer of dielectric has to be used to separate them, giving another extra amount of capacitance in the same space. The whole assembly is then encased in wax or plastics. A third basic type of capacitor uses a less easily understood principle, that of electrolytic polarisation, and is dealt with later under the heading of electrolytic capacitors.

The construction used for capacitors determines what voltage can safely be placed between the plates without risk of sparking across, what capacitance values can be achieved by the sizes of capacitors which will fit reasonably into circuits, and the behaviour at high frequencies and at high temperatures. We shall now examine in detail the various types of capacitors.

Practical Capacitors

Several types of capacitor are made simply as flat plates, usually by coating an insulating plate with metal on each side. A single plate cannot give a very great value of capacitance, so several plates can

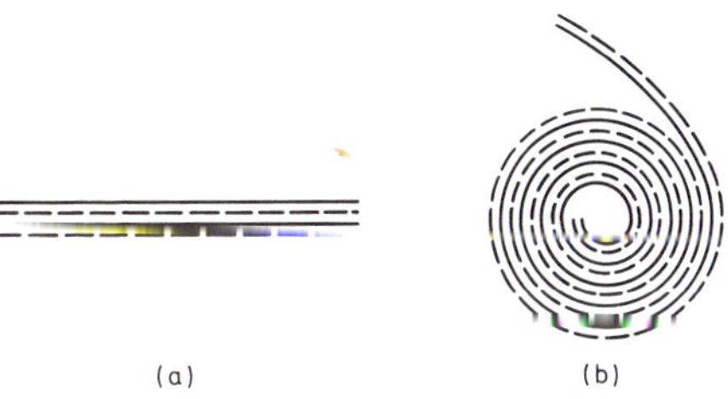

Fig. 3.4 (a) foil sandwich construction, (b) foil sandwich rolled up to form tubular capacitor.

be stacked together and connected in parallel for the higher capacitances up to about 10nF. In some cases where very thin insulators can be used at low voltages (up to 20V) capacitances of up to 0·5μF can be obtained. The capacitors made in this way usually use insulators of mica or ceramic, and are known as micas (silver-micas) or ceramic discs.

The actual material used has a great influence on the way in which the capacitor behaves, as we shall see later. Ceramic insulators are also used in the form of tubes which can be metal coated inside and outside to give tubular ceramic capacitors. In each case, a plastics moulding is applied round the finished capacitor.

Another family of capacitors uses the rolled-up sandwich construction described already. The insulator may be of paper which has been treated with wax to make it waterproof and so a better insulator; other waterproofing materials include silicones. Such paper-dielectric capacitors are still the most widely used group of capacitors. New plastics materials are now being used in the same type of construction, giving polyester, polystyrene, and polycarbonate capacitors.

The third form of the fixed capacitors (those in which the value cannot be deliberately varied while the capacitor is in use) is the electrolytic capacitor. The construction of the electrolytic capacitor is similar to that of a torch battery.

When a battery is flat, it is often because the conducting plates are covered with tiny bubbles of gas, usually hydrogen. Since the gas cannot conduct electricity the battery cannot deliver current, but the bubbles form a very thin layer of dielectric between the conducting plate and the conducting liquid or paste around it so that the whole arrangement has a large capacitance.

In a battery we try to stop this effect by adding a chemical which removes the gas. In an electrolytic capacitor we encourage the gas bubbles to form, making the conducting surfaces rough to retain the greatest possible amount of gas (greatest surface area) and removing any material which might attack the gas. The space between the conducting plates is filled, as in a battery, with a jelly or paste of chemicals which conduct electricity and give off gases whenever current flows. This paste is called an electrolyte and gives the type of capacitor its name.

Electrolytics

Electrolytic capacitors behave very differently from other types of capacitors, and they have had a reputation for poor reliability in the past caused to a large extent by a lack of understanding of their peculiarities. Although their reliability and other characteristics have been greatly improved over the past twenty years, it should be understood that a freshly made electrolytic has no capacitance at all, only a small resistance. Current must flow, causing the electrolyte to coat the plates with gas before the capacitance appears and the electrolytic starts to behave as a true capacitor, blocking d.c. and passing a.c.

If an electrolytic capacitor has been stored unused for some time (over a year) the gas film may have deteriorated and must be reformed. Sometimes this can be done within equipment when a capacitor is replaced and the current switched on. In some circuits, the surge of current through the capacitor can damage other components, and the electrolytic must be reformed before being replaced. Reforming is simply a matter of applying working d.c. voltage to the capacitor through a resistor large enough to stop more than a hundred mA or so passing, for a time of about a minute. The need for reforming is very much less with recently manufactured capacitors.

When an old piece of equipment, which uses electrolytics, is used for the first time in years it is very worthwhile to bring up the mains voltage slowly from an autotransformer of the *Variac* type to give the electrolytics a chance to reform before taking the full working voltage. The usual type of electrolytic must be connected one way only in a circuit, with the positive d.c. voltage on the indicated terminal.

Reversing the connections enables high currents to flow, and using a.c. alone, with no d.c. present, can also cause failure by breaking down the gas layer. Note incidentally that the resistance of electrolytics in series with the current can be quite high, and may restrict the use of some types. Old electrolytics can have series resistances of several hundred ohms, making them useless for coupling low impedance circuits.

Variables

The last group of capacitors is comprised of the variable capacitors. A variable capacitor consists of a set of fixed (stator) plates and a set of moving (rotor) plates which can either approach closer to the fixed plates or mesh with them to form different areas of overlap. The familiar radio tuning capacitor uses a set of plates mounted on a shaft which can be rotated into mesh with a set of non-moving blades. Maximum capacitance required in variables is usually 500pF, though variables formed of plates sandwiched by plastics sheets and varied in capacitance by varying the pressure forcing them together can have higher values.

The latter are known as compression variables, and they are usually used as trimmers. Trimmer capacitors are used in the same way as preset resistors, adjusted to give the desired capacitance and then left alone. Concentric trimmers use plates in the form of concentric cylinders, the fixed set of cylinders being meshed with the moving set which is screwed down inside the fixed set.

How Capacitor Types are Chosen

Capacitors may be used for transferring a signal, but not z.f., from one part of a circuit to another (coupling), for removing an unwanted signal without affecting d.c. voltages (decoupling); and for use with resistors and inductors to separate out signals of different frequency (filtering).

The factors which affect choice are the maximum operating frequency, working voltage, insulation resistance, power factor and temperature coefficient. These factors vary considerably with different types of capacitors and frequently make it impossible to interchange capacitor types. The serviceman must therefore know what is required of a capacitor used for a particular purpose so that a suitable replacement may be chosen.

Working Voltage

Since all capacitors are made of conductors separated only by a thin film of dielectric, at some high voltage between the conductors, sparking must take place. Before this actually happens, serious changes take place in the dielectric material causing it to become partially conducting. The dielectric is said to break down, and the capacitor fails.

Capacitor manufacturers set a limit, termed working voltage, for each type of capacitor which they manufacture, and they guarantee that failure due to breakdown does not take place if the capacitors are worked within this limit. Where a single working voltage is quoted, this is d.c. working voltage. An a.c. working voltage is sometimes quoted also, this is always less than half the d.c. working voltage, because a.c. volts are usually quoted as meter-measurements of r.m.s. volts and the true volts are peak-to-peak volts.

Where a.c. and d.c. conditions are present together, a wise rule is to take the working voltage as the d.c. volts plus three times the r.m.s. a.c. volts; for example, 150V d.c. and 50V a.c. gives 300V working. Working voltage is usually abbreviated as v.w. (volts, working).

Electrolytic capacitors have low working voltages (as low as 6V in electrolytics for small transistor amplifiers). The highest working voltages require ceramic or oil dielectrics (20kV working for capacitors in some industrial r.f. heating equipment). Between these two come paper dielectrics, used considerably in the range 250V–600V and polyester, which, though better insulators, have slightly lower working voltages due to problems of construction.

Mixed dielectric capacitors use a combination of paper and polyester to obtain high working voltage (due to the paper) along with very low leakage (due to the polyester) in one capacitor.

Maximum Operating Frequency

The reactance of a perfect capacitor should decrease steadily as the signal frequency rises. Every piece of wire has inductance, however, and coiled pieces of foil have higher inductance, so the perfect capacitor cannot be made; all real capacitors have some inductance and resistance just as all inductors have some capacitance and resistance and all resistors have some inductance and capacitance. The effect of this is that the reactance of any capacitor decreases as signal frequency is raised, but at some high frequency reactance starts to rise again.

The approximate frequency at which reactance stops dropping is the maximum operating frequency, and this can be measured for each type of capacitor. This frequency is approximate only, since the inductance of the leads to the capacitor is different in each circuit, but it serves as a good guide to the maximum frequency at which any capacitor type should be operated.

For block paper capacitors, the maximum operating frequency can be as low as 1·0MHz, for tubular paper and polyester 100MHz, for silver mica 100MHz upwards (depending on physical size—the smaller the dimensions, the higher the operating frequency) and for ceramics, 1000MHz upwards.

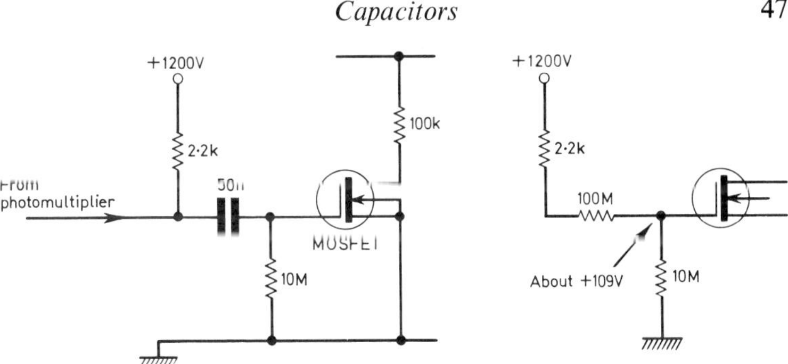

Fig. 3.5 Effect of a leaky capacitor. (a) circuit of MOSFET amplifier first stage for photomultiplier, (b) effect on bias if the capacitor has a leakage resistance of 100MΩ. This would destroy the first stage MOSFET.

Insulation Resistance

Electrolytic capacitors have very low maximum frequencies. Where electrolytics may have frequencies above 1kHz applied as well as lower frequencies, it is wise to use a paper capacitor in parallel. This technique is often used in transistor amplifiers of large bandwidth. Insulation resistance is frequently important when a capacitor is used to couple a.c. signals between different d.c. voltage levels.

As far as d.c. is concerned, a capacitor between two steady voltages acts as a resistor, this resistance being the insulation (or leakage) resistance. The combination of the insulation resistance and other resistance in the circuit forms a potential divider which alters the d.c. voltage present at the junction of capacitor and resistors. (Fig. 3.5).

Capacitors used to couple circuits of different d.c. voltages where resistors are of very high value, must be of very high insulation resistance. Examples of such circuits are coupling between valves and between MOSFETS, particularly in the special applications of electrometers, used to measure very small currents and voltages.

Low working voltage paper capacitors may have quite low insulation resistance (a 5μF block paper capacitor would, typically, have an insulation resistance of 40MΩ) and the higher the capacitance the lower the resistance. Tubulars have resistances exceeding 10,000MΩ. Mixed dielectrics have insulations better than 50,000MΩ, as also have silver mica types.

Ceramics generally, have the highest insulation (over

1,000,000MΩ) except for the types known as 'High Permittivity'. Such ceramics use a high permittivity dielectric to obtain a comparatively large capacitance in a small space, but the materials used have smaller insulation resistance, and the capacitance varies greatly with voltage and temperature. Electrolytic capacitors have low and variable insulation resistance.

Power Factor

Power factor is another way of expressing the lack of perfection of the capacitor. When volts are applied to a resistor, current flows and power is lost as heat, the power being equal to the volts × amps. In a perfect capacitor, when an a.c. signal is applied, there is a voltage across the capacitor and current through it, but no power is lost. In real capacitors, a very small amount of power is lost, equal to the volts × amps × power factor.

For low voltage block paper capacitors, this power factor may be about 0·02 at a frequency of 1kHz (100W a.c. applied, 2W lost in the capacitor) and large a.c. signals can cause serious overheating in the capacitor. For silver mica, the power factor is typically 0·001 and there is much less risk of heating. Power factors of electrolytics are high, 0·15 for higher voltage types to as high as 0·45 for low voltage types, so that overheating readily occurs.

When equipment is being 'soak-tested' (left running under observation for long periods) it is wise to strap a thermometer to each electrolytic (using aluminium foil as strapping material) to check running temperatures and to note these temperatures every fifteen minutes. Too rapid a rise indicates that more electrolytics in parallel must be used to share out the power dissipated.

Temperature Coefficient and Rating

The capacitance of paper and polyester capacitors rises as temperature increases (10,000pF capacitor increases by 2pF for every 1°C, (or 2°F) rise; a 20,000pF increases by 4pF per 1°C, and so on). In the application of paper and polyester capacitors this is usually unimportant, and the temperature range of working is of more interest. Temperature ranges of −40°C to +70°C are usual for paper dielectrics, −40°C to +125°C for polyester and mixed dielectrics.

Silver micas increase also in capacitance as temperature rises, but the increase is very small. In filtering circuits (including oscillator circuits) the change may be enough to cause trouble, and temperature compensation may have to be used. Ceramics may be made whose capacitance becomes less as temperature rises (NTC types).

Such ceramics can be used in parallel with other ceramics whose capacitance increases with temperature (PTC), or with silver micas, to form a combined capacitor whose capacitance does not change with temperature. The terms PTC and NTC have the same meaning as for resistors, that an increase of capacitance with increasing temperature is positive and vice versa. Ceramics and micas have working temperature ranges of −40°C to +100°C.

Electrolytics have large positive temperature coefficients; more important is the fact that the insulation resistance decreases greatly as the temperature increases. Maximum operating temperature is usually 70°C, often with some restriction on the current which may be passed at this temperature. Minimum temperature is about −20°C, as the paste electrolyte stops conducting when it freezes.

Failure of Capacitors

Capacitors can fail in the following ways:

(1)—Severe change of capacitance.
(2)—Internal breakdown causing low or zero insulation resistance.
(3)—Open circuit.

Open circuits can be readily detected, as no signal can pass through the capacitor, though this may not be obvious in a smoothing circuit except as an abnormal amount of hum. The easiest check is to connect (by clips, for example), another capacitor in parallel.

Complete breakdown to zero insulation causes the capacitor to behave as a piece of wire, passing a.c. and d.c. equally. This is usually discovered on test by the presence of d.c. where it is not expected; for example, a positive voltage on the grid of a valve or an excessive forward bias on the base of a transistor.

Complete breakdown in a capacitor used for power supply smoothing leads to repeated fuse blowing until the capacitor is disconnected. In filter circuits where no d.c. is present, breakdown is less likely, but can cause large changes in the frequency selected by the circuit. In such a case, replacement is the only way of checking quickly; note that only one lead need be unsoldered.

Change of capacitance has its greatest effect in filter circuits, causing a shift in the frequency of operation. Substitution is the quickest check. Small changes of insulation resistance are again most evident where d.c. is present.

Summary of Capacitor Applications

Electrolytics, which may be obtained in very large capacitance values, are used for:

(a) Removing a.c. from a d.c. supply (smoothing). The a.c. should be of a frequency less than 1kHz, otherwise it is better to use paper capacitors in parallel if necessary to obtain a high value of capacitance. The current flowing through the capacitor should not be enough to cause overheating (see under 'Power Factor'). Working voltage should be well within limits, allowing for the full voltage developed when no current is being used, and for switching surges. Temperature should be kept below 70°C.

(b) Coupling in low impedance circuits, notably in circuits using transistors and ICs. In this application, a low insulation resistance may cause changes in the d.c. conditions; for high frequency use, any electrolytics used should be by-passed by connecting paper or ceramic types in parallel.

(c) Decoupling at low frequencies.

Paper and mixed dielectric capacitors are used for:

(a) Coupling, except where very high values of capacitance are needed.

(b) Smoothing, in the form of block paper capacitors at high voltages where electrolytics are unusable. Voltage limitations are sometimes overcome by connecting capacitors in series. When this is done, high value resistors should be connected in parallel with each capacitor to equalise the charge on each capacitor as shown in Fig. 3.6. If this is done with electrolytics, the case must be insulated from surrounding metalwork.

(c) Filtering circuits, but avoiding frequencies near the maximum operating frequency.

(d) Decoupling at frequencies up to 10MHz.

Silver mica capacitors are used for:

(a) Coupling and filtering where precise values are required (close tolerance), and also low temperature coefficients.

(b) Tuned circuits.

Ceramic capacitors are widely used at frequencies over 10MHz for coupling, decoupling and filtering. Tolerances are not so close as for silver micas, and temperature coefficients are not so low. Negative temperature coefficient ceramics are available for temperature compensation. High permittivity ceramics are used for decoupling only; they provide large capacitance values (compared to other ceramics) in a small space, but capacitance varies considerably with temperature and voltage.

Further Notes on Capacitance and Capacitors

Stray capacitance plays an important part in electronics, usually undesirable. Any two objects have capacitance to each other if they are at different voltages. From this it follows that part of a circuit has some value of capacitance to every other part to which it is not directly connected. In most cases this capacitance is small, 1pF or so, and may be ignored, but it may at times be important, as when it limits the upper frequency range of an amplifier, and it may form an important part of the capacitance in a tuned circuit in high frequency amplifiers or oscillators.

In addition to this, transistors have stored charge which causes them to behave as if they had large values of stray capacitance across their terminals, so that the inputs and outputs of amplifiers have some capacitance (perhaps 10 to 30pF) to earth. In some cases this can cause difficulties, and the stray capacitance is 'removed' by a circuit device called bootstrapping, dealt with more fully in the companion volume to this book.

Non-Polarised Electrolytics

Non-polarised electrolytics are obtainable which need no d.c. voltage applied, or which can withstand d.c. voltages of either polarity. Electrolytics using tantalum foil instead of aluminium, and obtainable under a variety of trade names, have much higher insulation resistance and lower power factor (and can be used unpolarised) enabling them to take the place of paper capacitors for several purposes. Electrolytics are now available at capacitance values of up to 1F (1,000,000μF).

Note that for capacitors, and for electronic equipment generally, all of the maximum values may not be usable simultaneously. For

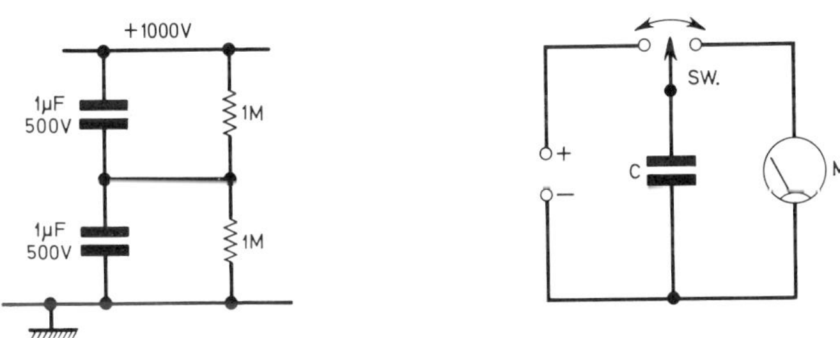

Fig. 3.6 (left) Using equalising resistors when capacitors are connected in series.

Fig. 3.7 (right) Principle of the charge-discharge capacitance meter.

example, a capacitor of quoted 500v.w., 2,000MΩ resistance, 70°C operation, 100MHz max. frequency may fail very rapidly in an application where all of these conditions exist at the same time. Generally speaking, capacitors made to military specifications will satisfy all of their maximum specifications together, but those made for domestic radio and TV will not, and should be derated.

Measurement of Capacitance

Rough measurements of capacitance values can be carried out using a meter working on the same principle as the ohmmeter, but using a.c., and a.c. meter. Such meters are useful for the larger values of capacitance only. A more modern type of meter uses the principle of charging the capacitor to be measured to a fixed voltage, and discharging it through a d.c. current meter.

If this is repeated at a rate above the rate at which the pointer of the meter can oscillate, the meter shows a steady reading. If the voltage for charging is V, and the capacitance value C, then the charge stored on the capacitor is $Q = C \times V$. If this is switched n times per second through a meter, the current is $n \times C \times V$, and can be measured.

A simple version of this meter uses a reed switch to connect the capacitor alternately to the battery and to the meter. This type of capacitance meter is simple and direct reading, and can be made sensitive at the cost of some sacrifice of simplicity (Fig. 3.7).

Capacitance can also be measured in bridge circuits similar to those used for resistance measurements, with the added advantage that the leakage resistance, and so the power factor, can also be measured.

CHAPTER FOUR

INDUCTORS

WHEN A CONDUCTING WIRE IS FORMED into a coil, its behaviour becomes noticeably different; it is now an inductor. Current flowing in a wire causes a magnetic field, and coiling the wire intensifies this effect greatly, causing the wire to act like a bar magnet when d.c. current is passed through the coil. Although a flow of current in a coil always creates a magnetic field, a magnet held in a coil of wire does not create an e.m.f. unless the magnet is moved; this is the principle of the dynamo.

In an inductor, both effects are involved. When a voltage is applied to an inductor, the current in the inductor changes from zero to some greater value which is fixed by the resistance of the inductor, and given by V/R. This current, during the time when it is changing causes a changing magnetic field, such as might be obtained from a moving magnet, and the changing magnetic field causes a voltage (e.m.f.) to be generated in the wire of the coil. What is more, the generated e.m.f. is in the opposite direction to the voltage applied, meaning that it is trying to push current in the opposite direction.

Current/Time Graph

While current is changing in an inductor, then, there is a little opposition, and as a result of this the current cannot change instantaneously. When a voltage is applied to a resistor, the current flows (practically) instantaneously, but when a voltage is applied to an inductor, the current increases slowly to its maximum given by V/R. The shape of the graph of current against time is called an exponential, and this shape keeps turning up in electronics. Fig. 4.1a shows an example.

The time taken to achieve maximum current (as near as we can measure) depends on the size and number of turns of the coil of wire, and increases enormously if the coil is wound on an iron (or similar) core. We define the inductance of the coil as the factor which causes the delay, and we keep the term 'self-inductance' for the case where a single coil is used.

If a rate of change of current of one ampere per second causes an opposing e.m.f. of one volt, the self-inductance of the coil is one Henry. Note that a rate-of-change of one amp. per second can be achieved by a change of 1/10 amp. in 1/10 second, the rate-of-change is

$$\frac{\text{change of current}}{\text{time taken for change}}.$$

The Henry is a fairly large unit, except for inductors used at mains and audio frequencies, and in electronics work the millihenry (mH) and microhenry (μH) are extensively used.

The A.C. Effect

The effect of the self-inductance on d.c. is a good clue to the a.c. effect. The voltage appears *before* the current, and, in an inductor where the self-inductance is high and resistance low, the voltage is $\frac{1}{4}$-cycle (90°) ahead of the current. Contrast the phase relations in capacitors. One good way of remembering the phases is the word C-I-V-I-L; 'in *C*apacitors, *I* before *V*; *V* before *I* in Inductors' (symbol *L*). Since the inductor tries to oppose changing current, the faster the change of current the more opposition (reactance). Since the voltage and the current are out of step, there is no power dissipation in a perfect (resistance free) inductor.

Inductors are not perfect, however. They are made of wire which has resistance and because of this the phase difference is never exactly 90° and there is always some power dissipated in the inductor, more than in a capacitor of the same reactance. Because of this, because the calculation of inductance is not so easy as that of capacitance, and because any form of magnetic core has a huge effect on the value of inductance, inductors are not so easy to work with as resistors and capacitors. Design involving inductors means a certain amount of cut-and-try work, since we cannot buy a range of preferred value inductors in the way we can buy capacitors and resistors. Fortunately, they are about the most reliable of components.

At frequencies below 200MHz, inductors are coils of wire. At higher frequencies, the wire need not even be coiled to have sufficient inductance, and at higher frequencies still the inductance of any piece of straight wire has too much reactance to allow easy current flow, so that we must send the current as a wave down a pipe (waveguide) or along lines where inductance and capacitance balance each other out (of which more later).

Inductors present rather a problem in modern circuit work where

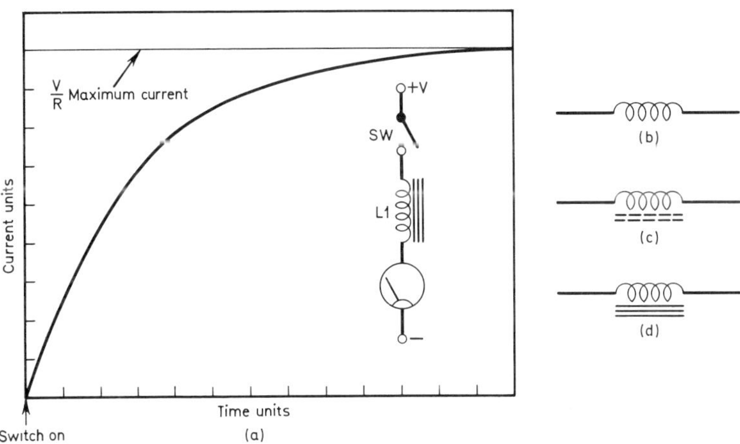

Fig. 4.1 (a) Exponential rise of current in an inductor, (b) Symbol for air-cored inductor. (b) symbol for dust-cored inductor (c) Symbol for laminated core.

the tendency is to form components and circuitry together. In film circuits, the inductor may be formed as a spiral of conductor if the inductance value required is low, or the inductor may be soldered on externally as in normal printed circuit practice.

Inductors can also be added externally to ICs, though the inductor is usually much larger than the IC; alternatively it may be possible to use circuits which enable capacitors and transistors to behave like inductors at signal frequencies.

Calculating Inductance and Inductive Reactance

Neither designers nor service engineers have to make up their own resistors or capacitors, but inductors have to be wound and adjusted. Large inductors should be bought in, even if it means a special order, but inductors of a few μH are easily made up and adjusted by cut-and-try methods. The easiest form of coil to work with, from the point of view of calculating inductance, is the one-layer solenoid, meaning a single layer of wire in a form where the length of winding is greater than the diameter (Fig. 4.2).

For small coils with no metallic or magnetic core, the inductance is given by the formula: $L = r^2 \times n^2/4s$, where L is the inductance value in μH, r is the coil radius in cm, s is the length of winding in cm, and n is the number of turns.

The usual problem, however, is to find the number of turns for a given value of inductance required, and this is always difficult because it requires a knowledge of the radius and length of the finished coil before it is wound. One useful tip is to use a set of formers of different radii and always wind so that the finished coil is 1cm long, different numbers of turns being achieved by using different gauges of wire or spacing out the turns.

If we take the ratio of length to radius (having fixed the length at 1cm), and call this ratio (now $1/r$) 'b', the formula can be rearranged to: $n = 5b\sqrt{L}$.

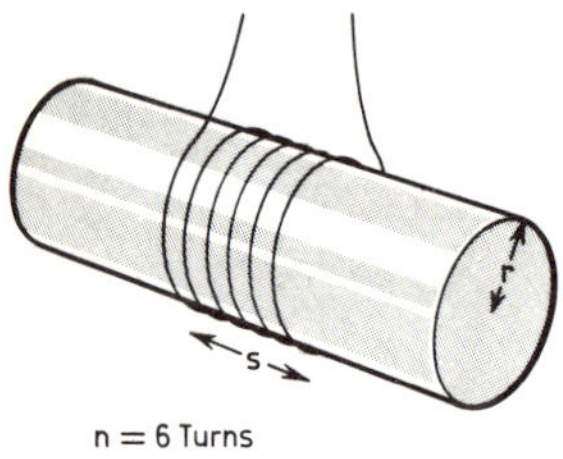

Fig. 4.2 The solenoid.

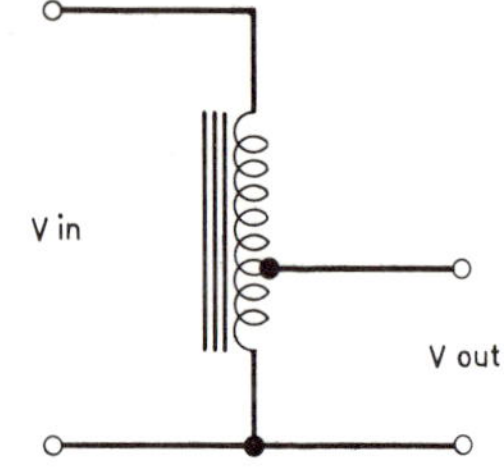

Fig. 4.3 The autotransformer.

A choice of formers of different radii makes it fairly easy to wind coils of any inductance value in the μH range. The ideal gauge of wire to use is one which gives the required number of turns with one wire diameter space between turns, and this can be found by using wire tables; for a coil wound to 1cm length, this means using a wire of diameter $1/2n$ cm, if the length of winding is s, the diameter is $s/2n$ cm. To make a well-spaced winding, wind on two wires together, and then remove one.

The reactance of an inductor is given by: $X_L = 6{\cdot}3\,fL$, where f is the frequency in Hz, and L is the inductance in H, giving X in ohms. If L is in mH and f in kHz, X is also in ohms, and if L is in μH and f in MHz, X is again in ohms.

The Effect of Cores on Coils

When a metal, metal dust, or oxide core is inserted into a coil, there is a change in the inductance. If the core material is non-magnetic (brass, for example) then the inductance is lowered; if the core material is magnetic (iron in any form) then the inductance is increased, often by a very great amount.

At high frequencies a solid iron core may have the same effect as a brass core, apparently to lower the inductance; this is because cores have two effects which cannot be separated in a solid core. The first effect is that all magnetic materials (more correctly, ferromagnetic materials) have the effect of concentrating magnetic fields. A coil with a core thus produces a more intense field within the core, which is just the effect which would be produced if we had increased the number of turns, and so it increases the inductance.

The factor which measures the increase in inductance is called 'relative permeability', and given the symbol μ, not to be confused with the μ in μH. If a coil has inductance of 1μH without a core, and 100μH with a core then the relative permeability of the core under the conditions of working is 100.

Permeability is not a constant. It varies according to the past history of the core (was it heated, taken near a magnet?) and the magnetic field in the coil. Manufacturers of cores have to provide graphs to show what the permeability of their cores will be under various conditions, and the graphs will be of no use if the cores have been ill-treated at any time.

The second effect of a core is 'eddy-currents'. We have seen that a varying current in a coil causes an opposing voltage to be induced. If a core, of any conducting metal, is inserted in a coil so that it shares the magnetic field some opposing voltage is induced in the core also, and currents circulate in the metal. This current opposes the coil current, but reduces the induced voltage in the coil, so making the inductance seem less, and wasting a fair amount of the signal power as well.

The permeability effect takes place only with ferromagnetic materials; the eddy current effect with all metals, but the eddy current effect is more effective at high frequencies. We can cut down eddy currents if we make it difficult for currents to flow in the core, which is why transformer and choke cores for the lower frequencies are made up of insulated laminations. At higher frequencies we get eddy current problems (and others) even in laminations, and cores must be made of iron dust embedded in plastics or of non-conducting materials based on iron.

Autotransformers

If a large inductor working at a low frequency is tapped by taking a third lead from a point which corresponds to, say, one tenth of the total number of turns, we should expect to find an alternating voltage between the tap point and the earth end on the inductor, and we would expect the voltage to be about one tenth of the voltage applied over the whole winding (Fig. 4.3).

This is so, and is exactly what we would expect from a tapped resistor. If we now draw current from this tapping, and measure the tap voltage, tap current and the main current as we increase the tap current drawn, we find that the tap voltage remains almost steady, but the increase in tap current is almost ten times the increase in the current drawn from the supply. Nothing like this ever happens in resistor networks. What is happening is that the current in the tapping is being supplied by the alternating magnetic field in the main winding.

When we draw one amp. out of the tap, we are reducing the magnetic field in the small portion of the inductor below the tap by a value corresponding to one amp. passing. That magnetic field is then made up by the rest of the winding, but, since the rest of the winding has ten times the number of turns, only about one tenth of an amp. is needed to make up the field to its old value.

Such an arrangement is called an *autotransformer*. When the tap can be varied in position it enables us to obtain any voltage between zero and mains voltage (or more), at a current limited only by the dissipation of the windings and the strength of magnetic field which the core can handle. The phase of the output is obviously the same as the phase of the input. The familiar *Variac* is of this form. Tapped coils are also used at higher frequencies. changing a high voltage low current signal (high impedance), into a low voltage high current signal (low impedance).

Mutual Inductance and the Transformer

If a second coil is wound over an existing inductor, then alternating current applied to the first coil causes an alternating voltage to appear on the second. If the coils are wound in the same direction, then the voltages across them are in phase, if one start of each winding is at a common voltage. In the simplest possible case, where there are no losses anywhere, the ratio of the voltages is the same as the ratio of the turns on the coils, and the ratio of the currents is the inverse of the ratio of the turns. So that: $V_1/V_2 = N_1/N_2$ and $I_1/I_2 = N_2/N_1$.

Although there appears to be a current gain or a voltage gain there cannot be both; there is no power gain. In real transformers (which are far from perfect) there is always a power loss, so that the voltage and current ratios are not quite equal to the turns ratio.

When two inductors share a magnetic field and behave in this way we say that they are coupled. We can measure how closely they are coupled by a quantity called mutual inductance. A mutual inductance of one Henry exists when a rate of change of one amp. per second in one coil causes one volt to be induced in the other coil. The definition is very similar to that for self-inductance, and the unit, the Henry, is the same.

At low frequencies it is difficult to obtain large values of mutual inductance. The magnetic field of the coil in which current is changing (the primary coil) must be arranged so that it all passes through the coil in which the voltage is to be induced (the secondary coil) at a low frequency. This means using a massive iron core to concentrate the magnetic field, and winding the coils as close together as possible,

At high frequencies the problem is often reversed. It is easy to ensure coupling between coils, difficult to avoid if it is not wanted. For this reason, coils used at radio frequencies are enclosed by a can if there are other coils near to which coupling must be avoided. The construction of the can and its spacing from the coil should be chosen so that there is not too much trouble from eddy currents.

If we wish to screen a coil from the effects of mains frequency induction from a large transformer, we must surround the coil with a material which has a large permeability, so that the magnetic field is concentrated in the shield rather than in the coil. Such a material is μ-metal, very often used for magnetic shielding in tape-heads and for the cathode-ray-tubes used in oscilloscopes.

Uses of Transformers

There are six types of service for which transformers are used in electronics: for voltage or current 'gain', for impedance matching, filtering, shaping, isolation or signal inversion.

Gain: When a transformer is used in a power supply, it may be supplying 30V at 10A for a transistor circuit or perhaps 10kV at 500mA for a transmitter valve. In each case it is being used to supply gain—current gain in the first place (as it would be wasteful to take 10A from the mains, using only 30V and wasting the remaining 200V in a resistor), and voltage gain in the second place, as we do not normally have 10kV mains available.

Matching: The gain use of transformers is often indistinguishable from their use for matching. Every circuit has an input impedance and an output impedance. These impedances, measured in ohms, are simple the ratio of signal volts/signal amps at the input and the output of the circuit respectively. A high impedance means that there is high voltage and low current, a low impedance that there is low voltage and high current.

When two circuits are connected together, there is always power lost in these impedances (as heat), but the loss is least when the impedances are matched. This means that the output impedance of the first circuit should be equal to the input impedance of the second to which it is connected. When the impedances are unequal, a transformer can be used as an impedance converter. If the impedances are Z_1 and Z_2, then the matching condition is: $N_1^2/N_2^2 = Z_1/Z_2$, where N_1 and N_2 are the number of turns on the transformer windings.

Example: match a 300Ω output impedance (a cable) to a 3Ω impedance (a loudspeaker). $N_1^2/N_2^2 = 300/3$, so that $N_1/N_2 = \sqrt{100} = 10$.

The transformer has a 10:1 turns ratio, the primary connected to the cable having ten times the number of turns of the secondary connected to the loudspeaker.

Filtering: A transformer can be tuned so that the signal transferred from primary to secondary is transferred most efficiently at one particular frequency. Methods of tuning are dealt with later.

Shaping: Chapter 1 outlined the ideas of wave shaping, and a transformer is one of the circuit devices which can carry out wave shaping. This is of particular use in pulse circuits.

Isolation: Some parts of a circuit are required to operate at a different voltage level from others. We are accustomed to the idea that the collector of a transistor is not at the same d.c. voltage as the base of the next transistor. In some cases the voltage difference may be very large. Preamplifiers for photocells may require a signal being transferred from a point at several kV to one only a volt

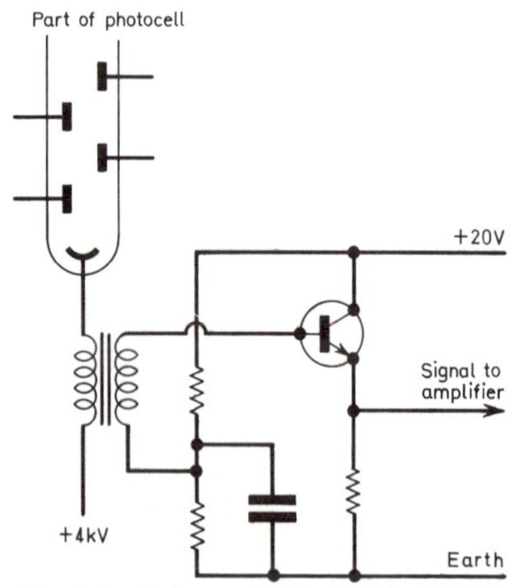

Fig. 4.4 Using a transformer for isolation.

above earth potential, A transformer may be used for coupling such circuits, especially when different 'earth' voltages are involved. An example is shown in Fig. 4.4.

Signal Inversion. At the secondary of any transformer, the signals at the two ends of the winding are mirror-images of each other. This effect is often (wrongly) called phase shifting by 180° and is much used in audio equipment such as public-address amplifiers. The èffect is also used in pulse circuits to select a positive or negative pulse from an input signal (See Fig. 4.5.).

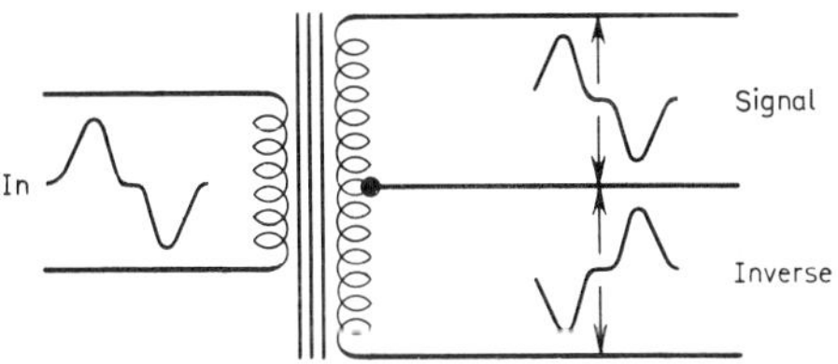

Fig. 4.5 Signal inversion with a transformer.

Transformer Cores

At very high frequencies, 100MHz or so, transformers are generally air-cored only. At lower frequencies, some sort of core is generally used to ensure that the magnetic field caused by the current flowing in the primary is sufficiently concentrated to couple closely to the secondary. The way in which the transformer behaves depends to a very great extent on the type and size of the core used.

Transformers apparently identical in size and number of turns may behave very differently if different core materials have been used. The most important features of cores for use in transformers or other inductors are the permeability at the conditions used, the hysteresis loss at the frequency to be used, the eddy current loss and the saturation field.

Permeability we have met already. This decides the final self inductance and mutual inductance of the windings. For low frequencies, the highest usable permeability is important to ensure high mutual inductance, and if the transformer output is to be a faithful replica of its input, the permeability must not vary too much as its primary current changes. At radio frequencies, high permeability is of much less importance, since coupling is more easily achieved, but the efficiency of coupling is still better when a fairly high permeability core is used. At v.h.f., the use of cores is undesirable.

Hysteresis loss is the work done, appearing as heat, when a material is magnetised and then demagnetised. With a.c. flowing in the primary of a transformer, this process is taking place all the time, and the energy converted into heat is lost. This means that the power obtained at the secondary is less than the power put in at the primary; the transformer is not perfect. The higher the frequency of operation, the greater the loss in this way because the magnetising and demagnetising processes take place more times per second.

Hysteresis loss can be lowered only by choosing core materials in which losses are low—unfortunately these are often materials in which permeability is low as well. The choice of a transformer core material, as with so many other things, is a matter of compromise.

Eddy Current losses. These have been dealt with earlier in this chapter.

Saturation field. At low currents in the primary winding of a transformer or any other inductor, each increase of current causes a large increase in the magnetic field in the core, much larger than would be obtained without a core. As current is increased, however, there comes a time when increases of current cause only very small increases of magnetic field. At this value of current, we say the core is saturated.

The amount of current in the winding which will saturate the core depends on the core size and material, and the number of turns in the winding. When a core is saturated, its inductance is very much lower than normal, and if this happens in a transformer then transformer action practically disappears and very high currents can flow in the primary winding. In addition, the output waveforms become severely distorted and the windings become overheated because of the excessive current in them.

Where transformers are used for signal currents only, saturation is seldom a difficulty, but where a large d.c. current flows as well, as in modulator transformers and in the output transformers of valve amplifiers, the cores may be working near to the region of saturation at times. In pulse work, the effects of saturation are often deliberately used to distort waveforms, and the design of magnetic amplifiers (see later) depends on the use of core saturation to control the inductance of windings.

Other Losses

Quite apart from the effects caused by the core, a transformer suffers from the effects of leakage inductance and of winding capacitance. Leakage inductance implies that not all of the magnetic field of the primary affects the secondary, so that some of the primary winding is simply behaving as a series inductor which has no effect on the voltage induced in the secondary.

The leakage inductance of a transformer can seriously affect its performance, and is particularly important if the transformer is to be used over a wide frequency range. Leakage inductance can be reduced by using a very high permeability core, by making the length of winding as great as possible, with as few layers as possible, by having the windings close together and by bifilar winding, in which the wires of the primary and secondary windings are wound together.

Winding capacitance is the stray capacitance which exists between points at different signal voltages on a transformer or any other inductor. It is of particular importance again at high frequencies, as it can resonate (see later) with the leakage inductance or with the winding inductance to make the response of the transformer very irregular at high frequencies. Winding stray capacitance can be reduced by increasing the insulation thickness on the wire, reducing the width of winding, increasing the number of layers and by keeping points with large signal differences well apart.

The requirements for low winding capacitance conflict almost completely with the requirements for low leakage inductance, and it is the transformer designer's headache to try to reconcile them and reach an acceptable compromise. Though computer-assisted design techniques help greatly, there is still a large amount of cut-and-try involved. The effects of winding capacitance and leakage inductance are shown in the equivalent circuit of transformers discussed in Chapter Six.

Note that, since the design of any transformer involves compromises in winding capacitance and leakage inductance, it is again undesirable to expect any two transformers to be interchangeable unless they have been made to the same design specification.

Servicing Notes

Inductors and transformers are not by nature troublesome components, and their failure is usually caused by trouble elsewhere, as when a power-pack transformer burns out due to a short-circuit on the output. Transformers which have windings of several thousands of turns of very fine wire (40 gauge or more) can give trouble due to electrolytic corrosion.

Water appears to penetrate the varnish coating of the wire, setting up a chemical action which is spurred on by the voltages present. This chemical action then eats away the wire, reducing its diameter until it fails, or until the heating at the thin portion causes failure. The writer has met a blocking oscillator in a TV receiver which had changed its primary resistance from 250Ω to 0·5MΩ (and was still working) because of such corrosion.

When a transformer has to be replaced, there may be some possibility of using a different type if it is a mains transformer but there is most unlikely to be any chance of using a different type at high frequencies, as explained earlier.

Since no transformer is perfect the voltage ratio is never quite as much as the turns ratio. Manufacturers quote only the voltage ratio as a rule, adjusting the turns ratio to give the quoted volts ratio. This means that a transformer ratio does not invert; for example, if a 3:1 voltage ratio needs a 4:1 turns ratio, then if the transformer is used the other way round the turns ratio will be 1:4 and the voltage ratio probably about 1:5 or less. For this reason, improvisations carried out by using the secondary of a transformer as its primary are seldom successful.

Mechanical Applications

Several devices use the magnetic effects of current in a coil to operate mechanical devices (relays, uniselectors). When mechanical work is done as well as electrical work (driving current) the mechanical and electrical effects cannot be separated, and the inductance of the coil depends on the work being done.

For example, an electric motor while working sets up an e.m.f. in the armature coil which opposes the applied voltage, so making the armature appear to have a very large inductance. If the motor is stalled (prevented from moving) the inductance of the armature coil appears to be very much less, as the only opposing voltage now comes from the inductance of the coil alone.

Core Gap

Large inductors wound on a laminated core would saturate the core if the core were a complete ring, allowing the magnetic field to build up. A gap is always left in the core to reduce the possibility of saturation, particularly if d.c. is flowing in the windings. The gap also reduces the inductance, and the size of gap is, as usual, a compromise between inductance and saturation field.

If an irreplacable transformer has to be rewound, very great care should be taken to measure the gap before taking the laminations

Fig. 4.6 Composite photograph of a large range of transformers and other inductors. Transformers in the background are for high voltage pulse work, hence the large insulators. (*Courtesy Gardner's Transformers Ltd.*)

apart, and to make sure that the same gap is used when the core is re-assembled.

Measurements on Inductors

Of all the common electronic measurements, those of inductance measurement are least often carried out except in the laboratories of factories specialising in inductors. There is no quick simple method corresponding to the capacitor charge/discharge meter described briefly in the last chapter. A form of direct-reading inductance meter can be set up, working on the ohmeter principle, but for large inductors only, and it cannot distinguish between reactance and resistance.

For smaller inductors, a bridge method must be used. As inductors are resistive, the power factor is usually much lower than in capacitors, and the bridge has to be balanced for power factor as well as for inductance. One simple design is the resonance bridge, which makes use of the fact that at one frequency (the frequency of resonance, see later), the reactance of an inductor is equal and opposite to that of a capacitor connected to it, and only the resistance of the inductor is left to obstruct current.

The bridge is adjusted to give a minimum reading on a meter or other form of indicator, whether by altering the frequency of the supply or by altering the value of the capacitor connected to the inductor. Then the resistance present in the windings is balanced out by a Wheatstone-bridge type of adjustment. The Maxwell Bridge circuit is also used for higher values of inductance.

For measuring mutual inductance, several bridges exist. The most useful is probably the mutual-inductance-capacitance bridge, where the signal transferred across the capacitance is balanced by the mutual inductance, giving again a minimum indication on the meter.

CHAPTER FIVE

NETWORKS I

A RESISTIVE NETWORK is a set of resistors connected together so as to produce an effect which cannot be produced by a single resistor. The failure of one resistor in a network can have serious effects on the operation of the network. The simplest and most often encountered network is the resistive potential divider. This network, as the name suggests, is used to obtain a potential lower than that of the supply with some degree of stability—meaning that the voltage obtained should not change too much if a small current is taken from it. The potential divider is found in bias supplies and in signal attenuators.

The action of the potential divider can be seen by referring to Fig. 5.1. By Ohm's Law, the current through the two resistors in series is: $V/(R_1 + R_2)$. The output potential above the 'low' end of R_2 is the voltage across R_2, and this is found, using Ohm's Law again, by multiplying the value of R_2 by the value of current flowing through it: $V_0 = V.R_2/(R_1 + R_2)$.

This equation is the general equation for finding the voltage output of a potential divider where no current is taken from the junction of R_1 and R_2. In the simplest case where the two resistors are of equal values, the voltage output V_0 is equal to half of the voltage input V.

The figure of V_0 obtained in this way is the true voltage only if no current is taken from the network, either by the transistor or valve being fed by the network or by a voltmeter being used to monitor the voltage. In some cases there may be no current, as when the divider supplies the grid of a valve, negatively biased, or the gate of a MOSFET. In other cases, the current drawn may be very small compared with the current $[V/(R_1 + R_2)]$ flowing through the network, as often happens in the biasing of the base of a transistor or the gate of an Enhancement Mode FET.

In both such cases, we can take the voltage output as given by the previous formula. If, however, the current taken from the network is not small compared to the network current, as might be the case with some power transistors or in the screen supplies of a pentode, then the formula becomes:

$$V_0 = \frac{R_2(V - i.R_1)}{R_1 + R_2}$$

Even in this case where the current supplied is comparable to the current in the network, there is some stabilisation of the voltage. As an example, take the supply to the screen grid of a pentode, as shown in Fig. 5.2. If the 200V supply were obtained by a series resistor of 10kΩ from the 300V supply, then a change of 1mA in the screen current causes a change of 10V in the screen voltage. With a divider network of 5kΩ and 20kΩ (nominal values—4·7kΩ and 18kΩ would be used), the voltage is again 200, but a change of 1mA causes only a 4V change in the voltage.

Units

In both of the formulae given, the units of resistance may be in ohms or in kΩ, providing both values of resistance are in the same units. For very high resistance potential dividers MΩ may be used, again as long as the same units are used for the two resistors. In the second formula, the units used for current must match the units used for resistance. If ohms are used, then the current must be in amps; this would probably be the case in the biasing of a large power transistor. If kΩ are used, the current should be in mA, and if MΩ are used, the current should be in units of μA.

Note that if either resistor of a potential divider fails, then the voltage changes. If R_1 goes o/c, then the voltage V_0 becomes zero (measured from the earth end of R_2), and this will also happen if R_2 goes s/c. On the other hand, if R_2 goes o/c, the voltage will rise to the value of the supply voltage if no current is taken from the network, but if current is being drawn, the rise will not be so great. If R_1 goes s/c, the voltage will reach the supply voltage, with the possibility of damage if, for example, the circuit is being used to supply the base of a power transistor.

Effects of High Frequency

Although the simple potential divider is entirely satisfactory at d.c. and at audio frequencies, it does not work so well at the higher frequencies encountered in video amplifiers and in the wide-band

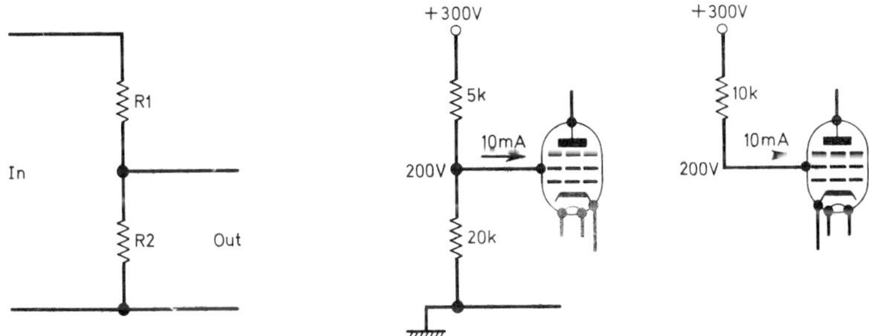

Fig. 5.1 (left) The resistive potential divider.

Fig. 5.2 (right) Current taken from the divider circuit.

amplifiers of oscilloscopes. For this reason, potentiometers, which are used as variable potential dividers, are not used to vary the signal gain in such circuits.

The cause of the trouble is the stray capacitance which exists between any two points at different voltages. Since the size of this stray capacitance depends on the physical dimensions of the resistor and its mountings rather than on its resistance value, the stray capacitance across each of the resistors in a divider is approximately the same, and at high frequencies, when the reactance of the capacitors is low compared with the resistance of the resistors, the capacitors take over the job of voltage division, and the signal voltage out is about half of the signal voltage in. Note that this affects signal voltage only, so that a divider which is used both for bias and for signal can have different division ratios for the two voltages present.

At lower frequencies, the situation is not quite so bad, but the effect is that the voltage division is not that set by the values of the resistors used, and it also varies with the frequency of the waveform applied in the frequency range where both resistors and capacitors affect the division. When fixed voltage divisions are used, there is a method of compensating for the effects of stray capacitance, but there is no way of compensating when a potentiometer is used.

For this reason, wideband amplifiers use sets of fixed voltage dividers, called attenuators, which are compensated for high frequency effects by the addition of fixed capacitors (Fig. 5.3). The method works by making the capacitor division ratio the same as the resistive division ratio, and by using capacitors which are large in value compared to the stray capacitance.

An extension of the simple potential divider is the dividing chain, used as an attenuator (Fig. 5.4). It offers a choice of output voltages,

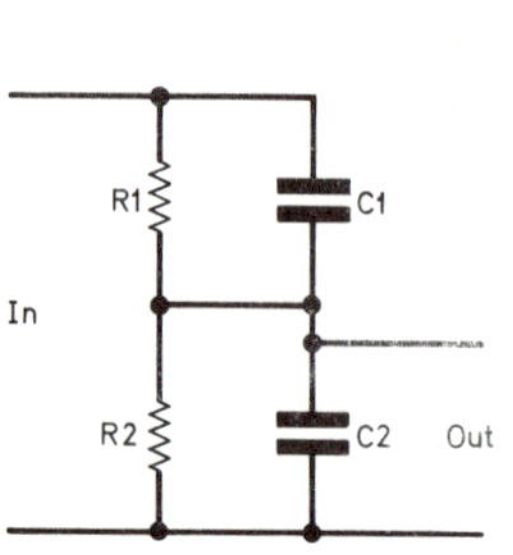

Fig. 5.3 Frequency compensated potential divider.

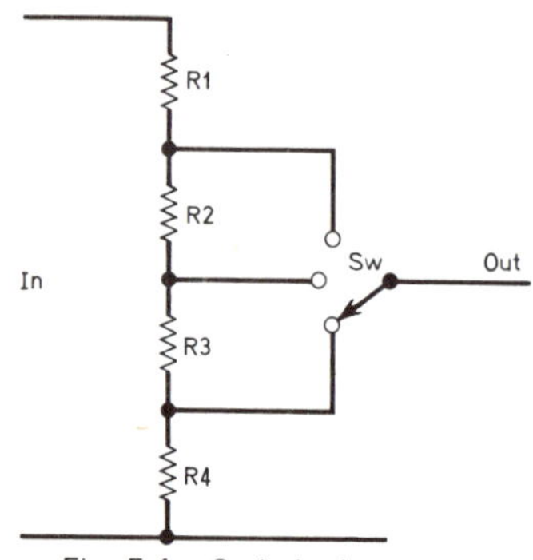

Fig. 5.4 Switched attentuator.

the value being found by treating all the resistors (in series) above the output point at R_1 in the simple divider, and all of the resistors (in series) below the output point as R_2.

The Bridge Circuit

For measurement purposes, the bridge circuit is probably one of the most important applications of the resistive potential divider. The straightforward Wheatstone Bridge circuit, named after its inventor, is shown in Fig. 5.5. Looking first at the left-hand side of the diamond shape, we can see that it is a voltage divider, so that the voltage output between the midpoint A and earth should be $V.R_2/(R_1 + R_2)$ if no current is being taken from the midpoint.

Similarly, if no current is taken from the potential divider on the right-hand side of the circuit, the voltage output at the midpoint B is: $V.R_4/(R_3 + R_4)$. When we use a bridge of this type, a current indicator is wired between the points A and B, and one of the resistors is adjusted in value until no current flow is indicated. Now zero current must mean that there is no voltage between A and B, so that: $R_2/(R_1 + R_2) = R_4/(R_3 + R_4)$. This is true if $R_1/R_2 = R_3/R_4$ or $R_1/R_3 = R_2/R_4$. In this condition the bridge is said to be *balanced.*

For measurement purposes, R_1 and R_2 can be a potentiometer with the resistance ratio marked on the dial, or a scale reading from 0 to 100. R_3 can then be a standard resistor of known value, or arrangements can be made to switch in various values of known resistors. The resistor which is to be measured is connected up in the position of R_4, and the potentiometer varied until zero current flows between the potentiometer tap and the junction of R_3 and R_4. When this happens, the ratio is read off the potentiometer dial, and the value of R_3 noted. From this, R_4 can be calculated.

For example, if the potentiometer reads 40 on a scale ranging from 0 to 100, and R_3 has the value of 10kΩ, then the value of R_4 will be $40/60 \times 10$kΩ, which is 6·66kΩ, indicating that the resistor was a close tolerance 6·8kΩ. Note that the potentiometer tap ratio is

$$\frac{\text{Reading on Potentiometer}}{\text{Full Scale—Reading}}$$

On a ten-turn potentiometer with a digital scale reading from 0 to 1,000, a reading of 374 would indicate a ratio of 374/626. All measuring bridges work on similar principles and follow similar circuit layouts.

Self-balancing bridges can be constructed in which the difference in voltage between A and B in the diagram of Fig. 5.5 can be amplified and used to adjust the tapping on the potentiometer until the bridge balances.

Other Uses for Bridges

Bridge networks are also used for other purposes. The arms of the bridge can be separated and fed from different voltages. When this is done, the point at which the bridge balances is decided by the ratio of voltages supplied as well as by the ratio of resistances. If we keep the resistance ratio fixed, this enables us to control a ratio of voltages, if one voltage can be adjusted until the bridge balances. In voltage stabiliser circuits, this is used to compare a voltage to be stabilised to a fixed voltage.

Bridge networks may also use resistors with different temperature coefficients, or may use thermistors in one or two arms. When this is done, the balance of the bridge may vary with temperature, and the bridge can be used for temperature measurement or compensation.

One other feature of a bridge network is often of great value. The

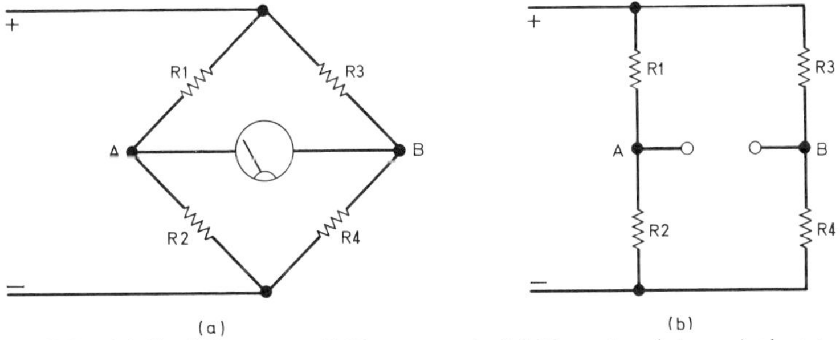

Fig. 5.5 (a) The Wheatstone Bridge network. (b) The network is equivalent to two potential dividers.

balance of a bridge is unaffected by the voltage supplied to it, and the positions of voltage supply and current detector can be interchanged. The bridge can therefore be supplied with a.c. without affecting its operation. Also, if we imagine a bridge circuit set up with junctions A, B, C, D; then, if a signal voltage is applied between points A and C, with D earthed, there will be no trace of the signal at B. Point B is often called a 'virtual earth' in this type of circuit, since there is no signal voltage or z.f. voltage between it and earth, yet it is not directly connected to earth (Fig. 5.6a).

It is even possible to feed another signal in at B, to mix with the first signal at A and C, while still preserving the condition that none of the signal between A and C reaches B. This is much used for signal mixing purposes, and also has valuable applications for 'earthing' a part of a circuit which cannot be directly earthed. (Fig. 5.6b).

Noise in Resistor Networks

In the section dealing with resistors, we saw that each resistor used in a circuit introduced a small amount of electrical noise. In most cases, the noise caused by the resistors is very small compared to the noise caused by the active elements in the circuit (transistors, etc.), but where a resistive network is used at the input of a high-gain amplifier, the resistor noise may be as large as the signal to be amplified, and low-noise resistors must be used for each resistor in the network which is not decoupled to earth.

In such networks, noise can also be kept to a minimum by the use of low values of resistance wherever possible, and by keeping the resistors as cool as possible. Low noise resistors in such a network should not be overheated during construction or servicing.

Measurements using Potentiometers

The potential-dividing action of a potentiometer is used for voltage measurements in the instrument correctly but confusingly called the potentiometer. If the voltage supplied across the potentiometer is known or, at least, stable, then an accurately calibrated potentiometer can be used to supply a known fraction of that voltage. A current indicator connected between the potentiometer tap and an unknown voltage will indicate zero when the potentiometer tap voltage equals the unknown voltage (Fig. 5.7). Since the potentiometer tap voltage can be read off, knowing the supply voltage, this enables the unknown voltage to be measured directly.

If the supply voltage is not known, then the potentiometer is first

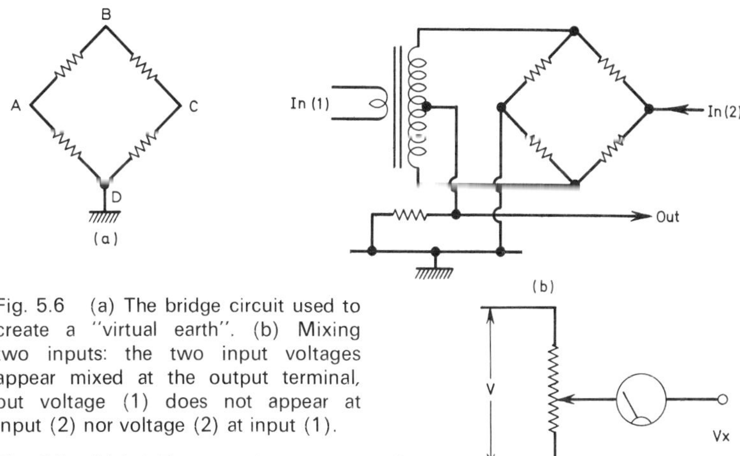

Fig. 5.6 (a) The bridge circuit used to create a "virtual earth". (b) Mixing two inputs: the two input voltages appear mixed at the output terminal, but voltage (1) does not appear at input (2) nor voltage (2) at input (1).

Fig. 5.7 (right) The potentiometer network.

calibrated by using a known voltage supply, such as a mercury cell, in place of the unknown and noting the reading. Other voltages can then be found by proportion. For example, if a mercury cell of voltage 1·406V balances at a reading of 86 on a 1,000 scale helipot, then a reading of 127 is a voltage of $1{\cdot}406 \times 127/86$, which is about 2·075V.

For voltage measurements, the potentiometer has the great advantage that practically no current is taken from the circuit whose voltage is being measured, at least when the potentiometer is balanced. This is of particular value in high resistance circuits, and the potentiometer may be made so that the voltage difference rather than current operates the balance indicator, so making the current drawn even less.

Potentiometer circuits are of particular value in measuring low voltages, and in measuring low values of resistance by passing a known current through them and using the potentiometer to measure the voltage drop. A bridge used for low resistance values has the disadvantage of measuring the resistance of any leads or connections as well; a potentiometer measures only a voltage between two points, and lead resistance does not affect the reading, because no current flows into or out of the potentiometer.

Capacitor and RC Networks

The simplest capacitor networks are potential dividing networks whose arrangement is exactly similar to the resistor potential dividers examined previously. Such circuits can be used to attenuate signals, though not passing d.c., in the same way as the resistive circuit, and they are independent of frequency as long as the reactance of the capacitors is low compared to the input resistance of the next circuit and the output resistance of the previous stage.

Some care has to be taken over the division ratio, however, as the reactance of a capacitor is proportional to $1/C$. Using reactances, we can write the division ratio for two capacitors as: $X_{C_2}/(X_{C_1} + X_{C_2})$. Because the reactance of a capacitor is $1/2\pi . f . C$, we can substitute this for the reactance values in the equation, and end up with the expression: $C_1/(C_1 + C_2)$.

The quantity on top of the expression above is the capacitance of the *upper* capacitor in the divider, and not of the *lower* as in the case of the resistive divider. Using similar working, we can prove that the total capacitance of a number of capacitors in parallel is given by: $C_{total} = C_1 + C_2 + C_3 + \ldots$ and the total capacitance of a number of capacitors in series is given by: $1/C_{total} = 1/C_1 + 1/C_2 + 1/C_3 \ldots$.

These expressions are just the opposite of the expressions for resistors in series and in parallel, the reason being that the reactance of a capacitor, which is the quantity corresponding to resistance, depends on $1/C$, and not on C.

Time Constant

Of all component networks, the most common are those which use resistors and capacitors together. One has already been described, the frequency compensated divider network of the early part of this chapter. In such a network, the condition for correct division at all frequencies was that the ratio of reactances should be the same as the ratio of resistances. In symbols, this is $R_1/R_2 = X_1/X_2$, but, since $X = 1/2\pi f . C$, this means that $R_1/R_2 = C_2/C_1$, or that $R_1 . C_1 = R_2 . C_2$. This last equation signifies that the product of R and the C in parallel should be constant for each pair of resistors and capacitors in the network.

This product of R times C is very important, because it measures the effect which a resistor–capacitor network has on a waveform; it is called time-constant. Although the combination of resistor and capacitor may be arranged in different ways, the product of $C . R$ still measures the effect.

When a capacitance in microfarads is multiplied by a resistance in megohms, the time-constant is in seconds. It may seem odd that the product of a resistance and a capacitance should be a time, but this is due to the way in which the units of resistance and capacitance

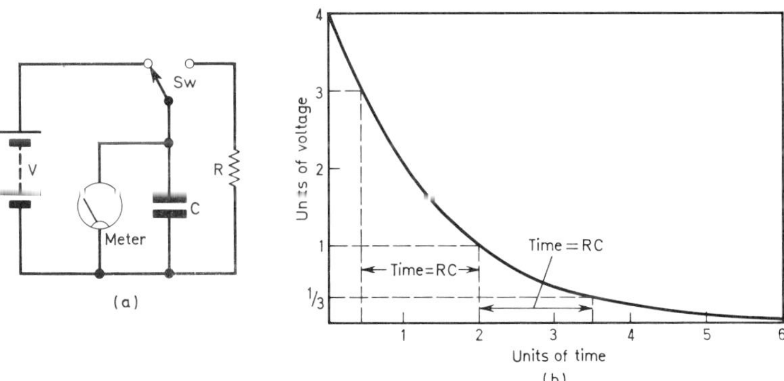

Fig. 5.8 Discharging a capacitor. (a) circuit, (b) graph of voltage across capacitor against time when capacitor is connected to resistor.

have been chosen. It is really no more odd than that the product of volts and amps should be the units of power, watts.

To understand the meaning of time constant, we must look at an example. In the circuit shown (Fig. 5.8), the capacitor can be charged to a set voltage and discharged by connecting a resistor across it. We must specify that the voltmeter connected across the capacitor takes no current from the capacitor, and that the discharging process is slow enough for us to follow. If the voltage is graphed every second, the shape of the graph is as shown. This shape of graph is obtained no matter what values of C and R are taken, though the scales of voltage and time may be different, and the shape is called an exponential curve.

The graph is the reason for the importance of the time constant, as the equation which describes the graph shape contains the time constant as its main feature. To see the effect for ourselves, however, take an example where the starting voltage is 90V and the time constant 10S. When the capacitor is discharging, the voltage drops to 30V in ten seconds (as near as makes no difference). It has dropped to one third of its starting value in one time constant of time. In the next ten seconds, the voltage has dropped to 10V, so that it has once again dropped to one third of its previous value in one time constant.

This is a general rule: whatever the voltage present across the capacitor when we start timing, it will drop to one third of that value in one time constant. In two time constants (20S in our example) the voltage will be one ninth ($1/3 \times 1/3$) of the starting voltage, in three time constants 1/27, and so on. A network using capacitors and resistors like this cannot allow rapid changes of voltage, the time constant decides how long the change must take.

The same idea of time constant applies if we start with the capacitor uncharged, and apply a voltage to the resistor and capacitor in series. In one time-constant time, the voltage across the capacitor rises until only 1/3 of the starting voltage is not across the capacitor (and 2/3 is); in two time-constants, 1/9 of the starting voltage is left (and 8/9 is across the capacitor), and so on. In each case the capacitor is trying to get to some final voltage (zero in the first case, applied voltage in the second) and the voltage it can reach in one time constant is 1/3 of the full voltage range away.

The curves of charge and discharge are of such importance that no more of this chapter should be read until the graph of voltage against time in each case is thoroughly understood. If equipment is available to plot the graphs practically (using a valve voltmeter, a 10μF paper capacitor, not electrolytic, and a set of resistors of 1 MΩ, 2·2MΩ, 3·3MΩ and 4·7MΩ), this should help the student greatly.

Remember that, when capacitance is in μF and resistance in MΩ, time constant is in seconds. If capacitance were in μF and resistance in kΩ, time constant would be in mS (millisecond = 1/1000 seconds) and capacitance in nF × resistance in kΩ gives time constant in μS (as does capacitance in pF and resistance in MΩ.

So far, we have taken the case of a voltage being applied to a capacitor through a resistor. What happens if the voltage is applied to the resistor through a capacitor? The answer is that the process is exactly the same since it is the same circuit, but in this case we are interested in measuring the voltage across the resistor, not across the capacitor. The circuit is shown (Fig. 5.9).

Starting at zero volts everywhere and switching on, the voltage across the capacitor does not change instantaneously (we know it takes one time constant to change so that 1/3 of the voltage is left). At switch-on, then, all the voltage appears across the resistor. As the capacitor charges up, more voltage appears across it, until, after

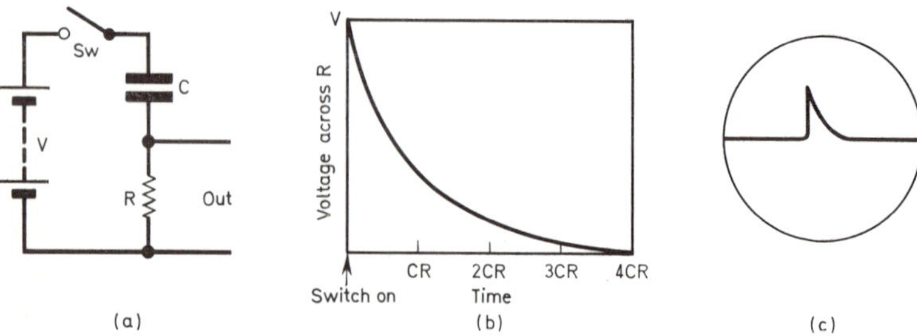

Fig. 5.9 Charging a capacitor through a resistor. (a) the circuit, (b) graph of voltage across resistor against time, (c) trace as seen on oscilloscope.

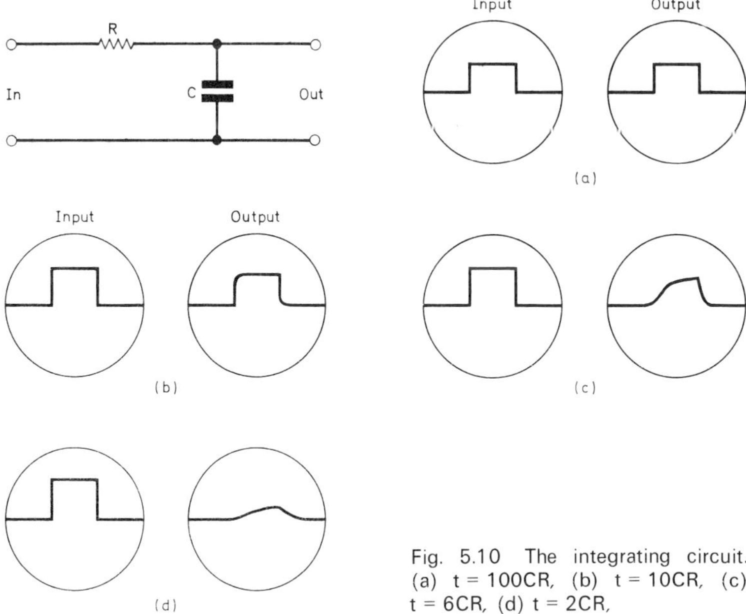

Fig. 5.10 The integrating circuit. (a) t = 100CR, (b) t = 10CR, (c) t = 6CR, (d) t = 2CR,

one time constant, only 1/3 of the battery voltage is left across the resistor. After two time constants, only 1/9 of the battery voltage is left across the resistor, and so on.

No matter how complicated a network may be, the effect of capacitors and resistors may be found whenever the supplied and final voltages are known. In the basic example shown, the voltages have been battery voltage or earth, but in more complicated networks several voltages may exist, with capacitors being charged or discharged to different levels. The important things to remember are that the capacitor charges or discharges between two levels of voltage and that the time to charge 2/3 of the way (so that 1/3 is left) is always the time constant.

Integrating Networks

We met the integrating network in Chapter 1 without learning of any circuit which could have such an effect. For many purposes, a circuit which gives roughly the correct response is quite suitable, and so the circuit shown is used as an integrating network or integrator. The effect which this simple network has on waveforms depends on the shape of the waveform and on the wave time used. The drawing of Fig. 5.10 should help to make this clear.

We have said earlier that all but 1/27 of an applied voltage has appeared across a capacitor in this network after 3 time constants. If we take this as being all the voltage (we can only be 1/27 out, which is better than 5% accurate (1/20), which in turn is more accurate than the values marked on most resistors or capacitors) we can draw each rise or fall of voltage as taking 3 time constants.

A square wave of 100 time constant in length would therefore look as Fig. 5.10a, one of 10 time constants as Fig. 5.10b, one of 6 time constants as Fig. 5.10c, one of 2 time constants as Fig. 5.10d. The shorter the wave time, the nearer the effect of the CR network comes to perfect integration of this waveform, and also the smaller is the voltage output of the network. Also shown is how the network affects the rise time of the waveform.

The effect of the integrating network on a sinewave is also of interest. Once again, if the wavetime of the sinewave is very much longer (100 times) than the time constant of the network, there is little effect. When a sinewave of shorter wavetime (higher frequency) is used, then the network causes a drop in amplitude and a shift of phase—the phase of the output sinewave lags behind the phase of the input sinewave.

When the wave time is 6·3CR, the reactance of the capacitor is equal to the resistance of the resistor, but, because of the phase difference, this does not make the network divide by a factor of 2.

Condition	Output V/Input V	dB Ratio	Phase Shift
R = Xc	$1/\sqrt{2}$ or 0·707	−3	−45°
R = 3/Xc or 1·73Xc	1/2 or 0·5	−6	−60°
R = 2·83Xc	1/3 or 0·33	−10·8	−70·5°
R = 3·87Xc	1/4 or 0·25	−12	−75·5°
R = 10Xc	1/10 or 0·1 (approx)	−20	−84·3°
R = 15Xc	1/15 or 0·068 approx	−23·4	−86·2°

Figures have been rounded off to the nearest whole number where possible. Note that for R = n/Xc, where n is a number greater than 15, Vo/Vin = 1/n and the phase shift is approximately 90°

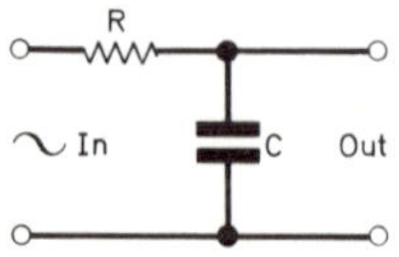

Fig. 5.11 The effect of an integrating circuit on a sinewave.

Instead, the output/input voltage is $1/\sqrt{2}$. In the language of sinewave amplitudes, this is '3dB. down'. At this same condition, the output sinewave lags behind the input wave by 45° (remembering that one cycle = 360°, 45° is 1/8 cycle). Fig. 5.11 summarises the amplitude and phase changes for a sinewave in the simple integrating network.

More complicated integrating networks can be made. If we put one simple integrator in series with another then, as we might expect, the integrating effect is increased—a squarewave becomes more like a triangular wave, with a greater reduction in amplitude; a sinewave has its amplitude reduced twice and its phase shifted further.

Three such networks can shift the phase of a sinewave by half a cycle if the time constants are suitably chosen; the time constant must be changed if the sinewave time is changed (Fig. 5.12). This is an example of a network which filters. It has the special effect of a phase change of 1/2 cycle on one particular wave-time; we shall see much more of this effect later.

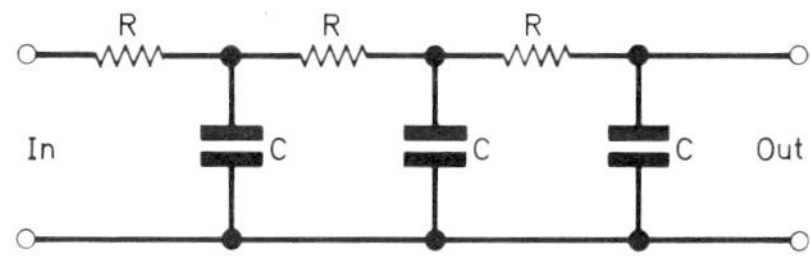

Fig. 5.12 Three-stage integrator or phase-shift network.

The integrating network is also found in power supplies, where a.c. is converted to d.c. and the remaining a.c. must be removed. An integrating network with a very long time constant is used to reduce the amplitude of the sinewave to a very low value (in the language of power supply designers, low ripple) so that the supply is very nearly pure d.c.

Differentiating Networks

If the simple CR network is turned upside down and the input fed into the capacitor the network now behaves as a differentiator for suitable wavetimes. If we use once again the idea that the effects of capacitor charging are over in 3 time constants, the wave shapes shown in Fig. 5.13 can be understood. If the wavetime is 100 times the CR time constant, the output is no more than a spike for each change of voltage.

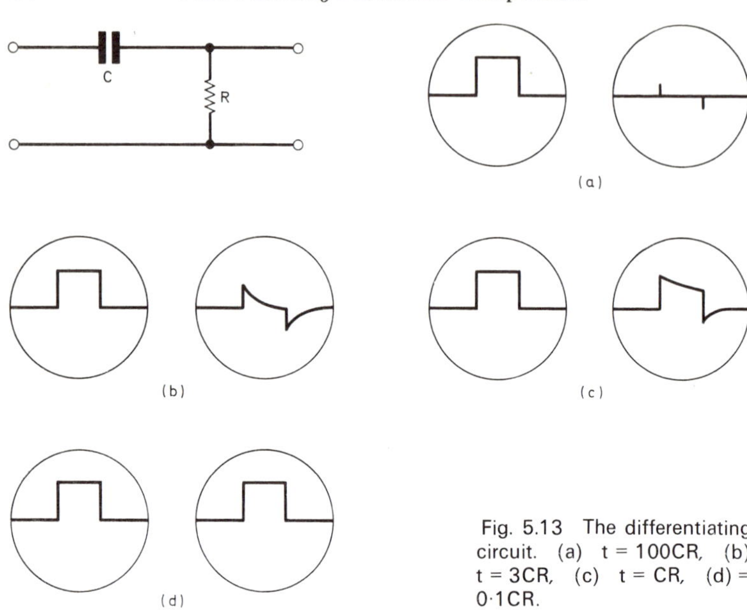

Fig. 5.13 The differentiating circuit. (a) t = 100CR, (b) t = 3CR, (c) t = CR, (d) = 0·1CR.

As the wave time becomes shorter, the effect of the network becomes less, and if the wavetime is so short compared to the time constant, that practically no charging of the capacitor takes place, the waveform is practically unaffected. A good rough approximation is to consider the waveform unaffected when the time constant of the differentiating network is more than ten times the wave time.

When the time constant is smaller than wave time, and the network is used for differentiation, the amplitude of the output depends on the rise time of the input waveform. If the input waveform has a rise time roughly ten times faster than the time constant, the amplitude of the differentiated output is practically the same as that of the input.

If the input waveform has a slow rise time (say ten times more than the time constant) the output amplitude is small, roughly one tenth (in the example) of the input. This is because the differentiating network passes only portions of waveforms with rise times of about its own time constant or less.

The differentiating network also affects sinewaves. If the wavetime is high compared to the time constant, there is little effect as in the case of a squarewave. As wavetime and time constant are made closer, the output amplitude drops and the phase shifts. Once again, when the resistance equals reactancc and wave time is $6{\cdot}3CR$, the amplitude is down by the factor:

$$\frac{\text{output}}{\text{input}} = \frac{1}{\sqrt{2}},$$

and the output sinewave leads the input by 45°. Fig. 5.14 summarises the amplitude and phase changes for a sinewave in the simple differentiating network.

As in the case of integrators, more complicated differentiating networks can be made by putting simple differentiators in series, and large phase changes in sinewaves can be obtained in this way.

The differentiating network is found as a coupling circuit in amplifiers. The time constant of the network determines the lowest frequency at which a reasonable amount of amplification can be obtained.

Estimating Effects of Networks

The design of networks is a complicated business, but a few simple rules can be used to predict roughly the effect which any given network has on a waveform. These rules are:

(1) During the rise and fall times of a pulse, capacitors may be imagined to act as resistors whose value is negligible, if the rise and fall times are much less than the time constant of capacitor and the resistor(s) connected to it.

(2) During the steady voltage periods of a pulse, capacitors act as

Condition	Output V/Input V	dB Ratio	Phase Shift
R = Xc	$1/\sqrt{2} = 0{\cdot}707$	−3	+45°
$R = Xc/\sqrt{3}$ (= 0·57Xc)	1/2 or 0·5	−6	+60°
R = 0·354Xc	1/3 or 0·33	−10·8	+70·5°
R = 0·26Xc	1/4 or 0·25	−12	+75·5°
R = 0·1Xc	1/10 or 0·1 approx.	−20	+84·3°
R = 0·068Xc	1/15 or 0·068 approx	−23·4	+86·2°

Figures have been rounded off to nearest whole numbers where possible. Note that where R = Xc/n, where n is a number greater than 15, Vo/Vin is approx. 1/n, and the phase shift is about 90°.

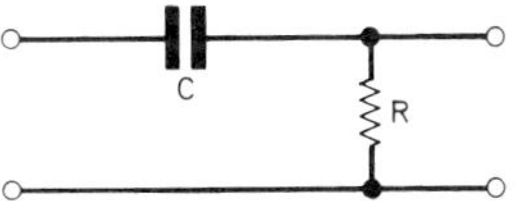

Fig. 5.14 The effect of a differentiating circuit on a sinewave.

a voltage supply which drops by 2/3 of original value in one time constant.

(3) Capacitors connected together act as dividers.

(4) The d.c. voltage at the output before and after the input waveform is the d.c. voltage connected to the resistor network at the output.

An example is shown. C_1 and C_2 act as a divider, and C_1R_1 is a differentiating network whose time constant is short compared to the input wavetime. The first step is to apply rule 3 and imagine the input amplitude decreased in the ratio $C_1/(C_1 + C_2)$. C_2 can now be ignored; we can start again with the decreased input at C_1R_1. This is differentiated as shown in Fig. 5.15. This waveform then passes to R_2C_3 which is an integrating network of time constant equal to the original wave. This tries to charge up to 2/3 of applied voltage in one time constant, but cannot get there due to the small wavetime of the waveform; the result is shown. Finally, R_3C_4 has a large time constant and therefore has little effect on the waveform; it is purely for coupling. Starting and finishing voltages are -10V, since this is the connection to R_3.

Failure of either resistor or capacitor in a CR network causes large changes in the effect on a waveform. If either capacitor or resistor changes to a lower value, the time constant of the network becomes less; if either resistor or capacitor changes to a higher value, the time constant becomes more, and the effect of the time constant on the waveform is as described previously.

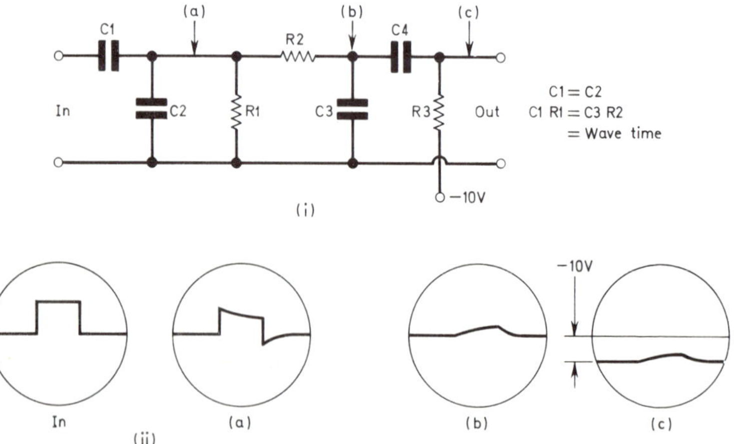

Fig. 5.15 Effect of a complex network on a waveform. (i) the network, (ii) oscilloscope traces at the lettered points.

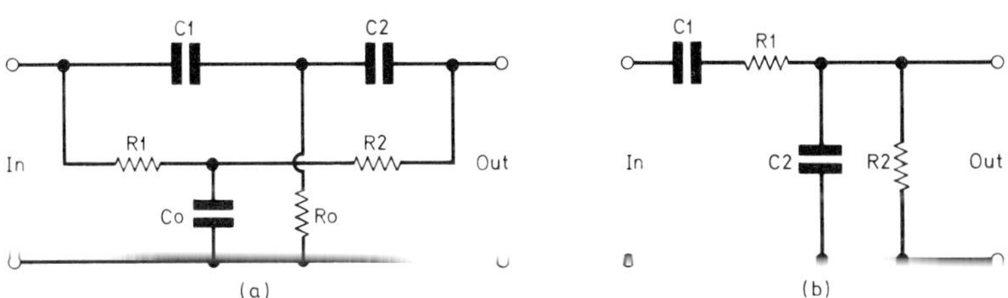

Fig. 5.16 (a) parallel-T network, (b) Wein bridge network.

If a resistor becomes s/c in an integrator, the effect on the waveform is to produce less integration, but this may not be obvious in a three stage integrator. The same sort of failure in a differentiating network would short out all the signal. Similarly, a s/c capacitor in a differentiating network would cause less or no differentiation, but this fault in an integrator would short out the signal.

Advanced Notes

Phase shift networks: Triple integrators or differentiators as shown are used in low frequency generators (oscillators); frequently some of the values may be variable. The effect of each network is to shift some frequency of sinewave by exactly 180° (1/2 cycle) and the frequency of the sinewave may be calculated from the time constants used.

Parallel-T networks: These are frequency-selective networks (see Fig. 5.16a) which reject the frequency given by

$$f = 1/(6{\cdot}3 \times C_1 \times R_2),$$

when $R_1 = R_2 = 2R_0$ and $C_1 = C_2 = 1/2C_0$.

Wein bridge: This network (shown in Fig. 5.16b) passes the frequency given by $f = 1/(6{\cdot}3 \times C_1 \times R_1)$, when $C_1 = C_2$ and $R_1 = R_2$.

Both the Wein bridge and the Parallel-T (along with triple integrators or differentiators) are used in oscillators for frequencies ranging from near z.f. to about 100 kHz. In such use, the networks are placed in the feedback circuits attached to amplifiers.

CHAPTER SIX

NETWORKS II

THE DECIBEL SCALE HAS ITS ORIGINS well before the rise to importance of electronics. It was noticed by early psychologists working on the senses of hearing and sight that the human response to sound and light was not linear. Putting it another way, doubling the amplitude or a sound or the luminance of a light source did not double the loudness heard by the ear or the brightness seen by the eye. We have found that the response of the ear or eye is best fitted to a logarithmic scale, where the effect is measured by the logarithm of the power output of sound or light. Several such logarithmic units have been devised, but the one which was most used in telephone engineering has been adapted into electronics as the decibel.

The Decibel Scale

The decibel scale is used to compare power levels at different points, and is defined by: No. of dB $= 10 \log P_1/P_2$ where P_1 and P_2 are the two levels of power. For example, if P_1 is ten times P_2, then the ratio in dB is 10 log 10, which is 10dB; if the ratio were 100:1, then the dB ratio would be 10 log 100, which is 20dB. The decibel is one tenth of a larger unit, the Bel, now seldom used.

Because the decibel scale represents ratios, it makes no sense to speak of a sound having a level of, say, 90dB, unless we know to what power this is being compared. In much acoustical work, the reference level is an intensity of sound of 10^{-16} watts per cm^2, so that 90dB represents about 10^{-7} watts per cm^2 intensity. In electrical work, the reference figure is usually 1mW, so that a 90dB output would represent the huge output of one million watts. For this reason, it is important to find out what reference level is being used whenever the dB scale is used.

Further confusion is caused in electrical engineering by the common use of the decibel scale to express current and voltage ratios. This can be justified if the comparison is made between points having the same impedance, because the power levels are then, in voltage terms, V_1^2/R and V_2^2/R, if V_1 is the voltage signal at the input, V_2 is the voltage signal at the output, and R is the impedance at each end of the network. Similarly, if we measure current signals, the power levels are I_1^2R and I_2^2R.

Because the ratio of powers is now V_2^2/V_1^2 or I_2^2/I_1^2 we can write the ratio in dB as 10 log V_2^2/V_1^2, which is 20 log V_2/V_1, or as 10 log I_2^2/I_1^2, which is 20 log I_2/I_1, because squaring a quantity can be carried out by doubling the value of the logarithm. These expressions are widely used for expressing voltage and current ratios in dB (see Table 6.1).

Power Ratio	Voltage or Current ratio	dB
1·12	1·06	0·5
1·26	1·12	1·0
1·58	1·26	2·0
1·99	1·41	3·0
3·16	1·78	5·0
10·0	3·16	10·0
100	10·0	20
1,000	31·62	30
10,000	100	40
10^5	316·2	50
10^{10}	10^5	100

Table 6.1—DECIBEL SCALES

Though this type of use is convenient, it is not always justified, and care should be taken when voltages or currents at points with very different impedance values are compared, as the dB figure can then be used only as a guide to that particular network, but not as a comparison to any other network. For example, it is reasonable to use the dB scale to compare the loss or gain of a network with the loss or gain at other frequencies; it would not be reasonable to use the dB scale to compare the loss or gain of different networks with different impedance levels.

Inductive Networks

Networks of inductance can be used for differentiation and integration in the same way as the network of capacitance and resistance illustrated in the previous chapter; the important difference is that when inductance is substituted for capacitance in a network, the effect is reversed (a capacitive integrator becomes an inductive differentiator, and the other way round), and the time constant is L/R. For sinewaves, the impedance of an inductor L in series with a resistor R is given by $\sqrt{(R^2 + \omega^2 . L^2)}$, where ω is $2\pi \times$ frequency of sinewave. The phase angle between current and voltage across the network is $\tan^{-1}(\omega L/R)$, meaning the angle whose tangent is $\omega L/R$, the value of tangents being obtained from standard mathematical tables. An inductor in parallel with a resistor has an impedance of $\omega . L . R/(R^2 \times \omega^2 . L^2)$ and a phase angle of $\tan^{-1}$ $(R/\omega L)$.

Inductance-resistive networks are seldom used for differentiation or integration unless the inductor is already needed for another part of the circuitry, since inductors must be wound to order, and capacitors are available readymade. Where transformers are used to couple circuits, inductive-resistance networks frequently exist and are used to shape waveforms; conversely the use of transformers to couple circuits affects the waveform coupled, and the time constants must be suitably chosen if the waveform is to be unchanged.

Inductors and Capacitors

The networks using inductors and capacitors are of vital importance in electronic engineering. Take first the simple series circuit of a capacitor and an inductor across an a.c. signal as in Fig. 6.1a. Since the reactance of an inductor increases from practically zero as the frequency of the signal is raised from d.c., and the reactance of a capacitor decreases from a very high value at d.c., there must be some frequency at which the two are equal. At this point we must revise the effects of inductors and capacitors on a signal.

In an inductor, the voltage led the current by 90°, in the capacitor the voltage lagged 90° behind the current, and the effect of this was to make the voltage across the capacitor always opposite in direction to the voltage across the inductor. We describe this mathematically by making the reactance of an inductor a positive quantity and the reactance of a capacitor a negative quantity. The impedance of an inductor in series with a capacitor is therefore $(\omega L - 1/\omega C)$, and the impedance when the two are in parallel is $\omega L/(1 - \omega^2 LC)$. In each case, the phase angle is 180° between inductor and capacitor.

When the reactance of the capacitor is equal to the reactance of the inductor, the voltage across the capacitor cancels out the voltage across the inductor, though each of the voltages separately exists. As far as the series combination is concerned, there is no reactance. At the frequency where capacitive reactance equals inductive reactance, the only total impedance is the resistance of the

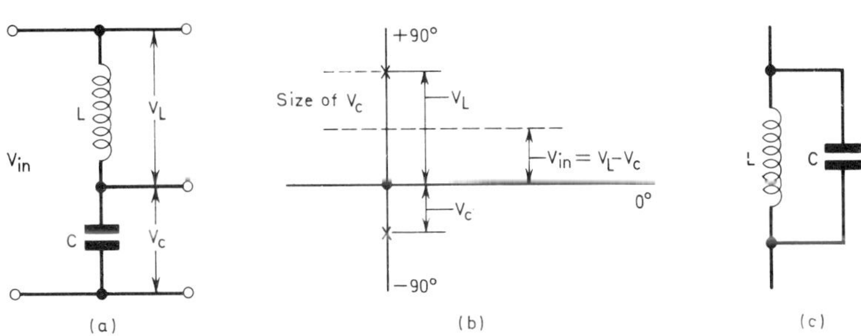

Fig. 6.1 (a) series L-C circuit. The same current flows through both components, and the voltages are 180° out of phase; (b) the phase diagram indicates the 180° phase difference. If V_C equals V_L, then there is no V_{in}, as there is no reactance, resistance or impedance; (c) parallel L-C circuit. The voltage across each is the same and it is the *currents* which are 180° out of phase.

wire used in winding the inductor and any resistance due to the capacitor. As both of these resistances are small, the current through the network will be very large if the a.c. source can supply a large current.

This will take place at the frequency given by: $f^2 = 1/4\pi^2 LC$ or approximately $f^2 = 25{,}000/LC$. In the first expression, f is in Hz, L in henries and C in farads. In the second expression, f is in kHz, L in mH and C in pF, or f in MHz, L in μH and C in pF. For example, if $L = 20$mH, $C = 50$pF, then $LC = 1{,}000$ and $f^2 = 25{,}000/1{,}000 = 25$, so that $f = 5$kHz. This effect of equal and opposite reactance cancelling out is called *resonance*.

The *Q* Figure

In addition, we must not forget that, though the voltages across the capacitor and across the inductor cancel when we add them together (by measuring across both components), they both still exist. Because the current through the two components is high, the voltage, given by $V = Z \,.\, I$, across either the capacitor or the inductor is also high. If the output of the circuit is taken across the capacitor or across the inductor, and the input is across both, then there is gain of voltage at the frequency of resonance.

This is not something for nothing, for there is no power gain; the current at the input is high, but very little current can be taken from the output, and the effect works only at, or very near, the frequency of resonance. The effect of changing frequency upon the 'gain' is shown in Fig. 6.2, where the shape of the curve is measured by the quantity called '*Q*'.

The factor, Q, sometimes called the quality factor, is:

$$\frac{\text{Reactance of capacitor or inductor}}{\text{Total resistance}}$$

when the network is at resonance. Q may be as low as 5 where a very large inductor is used at low frequencies, or as high as 300 for an inductor wound with stranded wire and used at high frequencies. The Q figure gives an idea of the step-up of voltage which can be expected, and also can be used to calculate the effect of frequency changes, as the gain is roughly 3dB down (0·707 of its maximum value) at a frequency given by $f/2Q$ away from f, the frequency of resonance.

For example: a series circuit resonates at 100kHz with a Q of 200. The gain is therefore 3dB down at $100/2 \times 200$kHz away from 100kHz. Note that the figure of frequency used was in kHz, so that the answer is in kHz, in this case 0·25kHz, or 250Hz. This means that the gain would be 3dB down on the 100kHz gain at 99,750Hz and at 100,250Hz. This would be termed a sharply tuned circuit with a narrow bandwidth. The bandwidth is the spread of frequency between two limits of gain, usually 3dB, and in this case, the bandwidth to the 3dB points is 500Hz (250Hz each way).

Series Tuned Circuit

This type of circuit is called a series tuned circuit and is found extensively in any equipment which must amplify signals of one frequency rather than any other; examples include sound broadcast

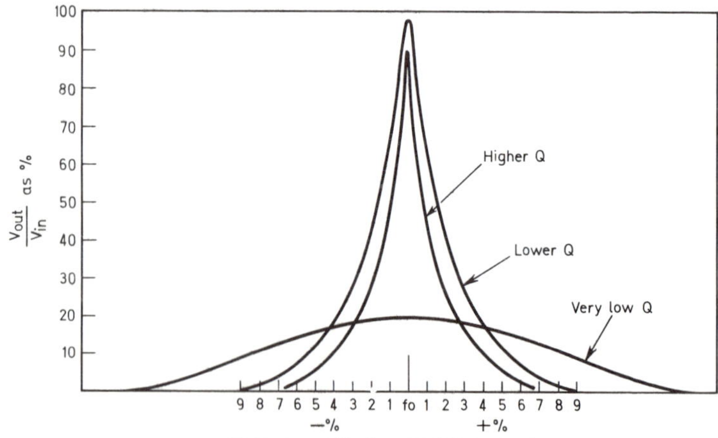

Fig. 6.2 Graph of gain against frequency near the point of resonance.

and TV receivers, radar receivers, all types of transmitters. The complete expression for the impedance of a series circuit containing inductance, capacitance and resistance is: $Z = R^2(\omega L - \omega^1 C)$ and the phase angle between current and voltage is:

$$\tan^{-1} \frac{\omega L - 1/\omega C}{R}$$

Note that in this case the phase angle depends greatly on ωL and $1/\omega C$. At resonance, current and voltage are in phase (an angle of 0°), but at a small frequency distance away from resonance, the value of $(\omega L - 1/\omega C)$ may be quite large and either positive or negative. A positive phase angle means that the network still behaves like an inductance, with V leading I; a negative angle means that the network behaves more like a capacitance, with I leading V.

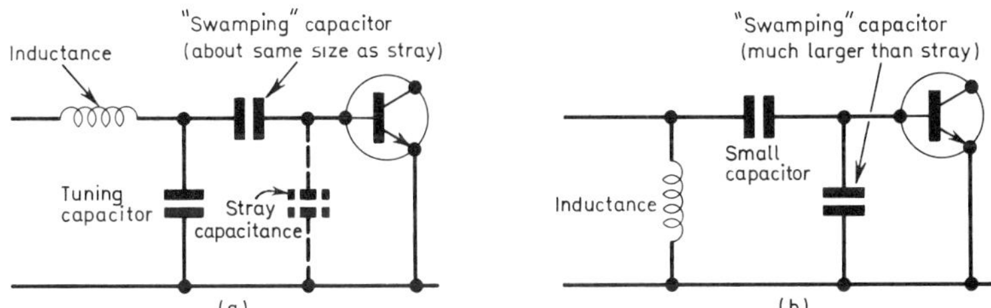

Fig. 6.3 (left) Swamping stray capacitance with a small C in series in a series-tuned circuit.

Fig. 6.4 (right) with a large parallel C in a parallel-tuned circuit.

Sometimes the inductance in a series-tuned circuit is a visible component but the capacitance is not, being formed by the stray capacitance between the base and emitter of a transistor or between the grid and cathode of a valve. In such cases, any change in the stray capacitance will cause a change in the frequency of resonance, and it is better to make the contribution of the stray capacitance negligible by adding a capacitor in series with it, using another capacitor across both to bring the combined capacitance up to the amount needed for tuning.

When the tuning is by stray capacitance or by very small capacitance values, (Fig. 6.3 and 6.4), tuning alterations are most easily carried out by altering the inductance of the inductor, using a variable position dust core. In the lower frequency circuits, where larger values can be used, it may be easier to vary the capacitance.

The Parallel LC Circuit

The parallel circuit is shown in Fig. 6.5. Again there is a frequency of resonance when the reactance of the inductor equals the reactance of the capacitor, but, because the components are now in parallel, the impedance of the network at resonance is very high. The impedance at resonance is roughly Q times the impedance at any frequency well away from the resonant frequency (perhaps about three times or a third of the resonant frequency), where Q has the same meaning as before. In addition, the variation of impedance with frequency near the frequency of resonance follows the same shape of curve as that shown for the series resonant circuit, and the same equation for the bandwidth applies.

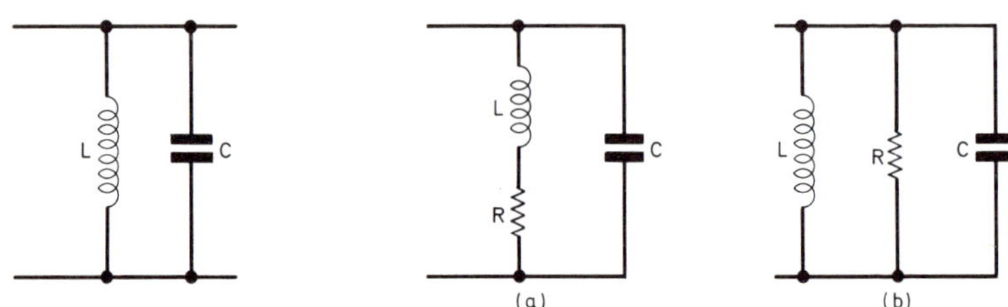

Fig. 6.5 (left) The parallel L-C circuit.
Fig. 6.6 (right) L-C circuits with resistance (a) where the resistance is fairly large, compared with the reactance, and is contained in the inductor, (b) where the resistance is small and is in parallel, as in many wideband amplifiers.

The complete expression for the impedance of a parallel resonant circuit shows two possible cases. The more common one is where the resistance is contained in the inductance, and so the circuit can be represented by a capacitance in parallel with the series combination of inductance and resistance as in Fig. 6.6. In this case the impedance is:

$$Z = \frac{R^2 + \omega^2 L^2}{(1 - \omega^2 LC)^2 + \omega^2 C^2 R^2}$$

and the phase angle is:

$$\tan^{-1} \frac{\omega[L(1 - \omega^2 LC) - CR^2]}{R}$$

In this case, the condition for resonance is not the familiar $\omega^2 = 1/LC$ but $\omega^2 LC = 1 - \omega CR$, which makes the ω value for resonance slightly different from that obtained if no resistance is present. In many cases, we can neglect the resistance. When the resistance of the inductor is negligible, and there is a resistor wired in parallel with the network, the impedance is

$$Z = \frac{1}{\sqrt{[(1/R)^2 + (\omega C - 1/\omega L)^2]}}$$

which gives the usual resonance condition of $\omega^2 = 1/LC$ and the phase angle is $\tan^{-1} R(1/\omega L - \omega C)$.

Choice of Series or Parallel Resonance

Since both types of tuned circuits appear to do much the same sort of thing, why should a designer use one in preference to the other? The answer to this lies in the impedances of the circuits connected to the tuned circuits. For example, if the circuit providing the signal to the tuned circuit had a very high impedance, it would make very little sense to connect to it a network which had a very low impedance at resonance, as the current flowing in the tuned circuit would be set more by the high impedance than by the tuned circuit.

It would, however, make sense to use a parallel tuned circuit, with a high impedance at resonance matching the high impedance of the circuit and causing a good transfer of power. At other frequencies, the impedance of the parallel tuned circuit is low and the transfer of power is very small. This is often the case at the outputs of active devices. At the inputs of transistors and some other amplifiers the impedances are lower, and a series tuned circuit is more suitable.

Coupled Tuned Circuits

If two tuned circuits are so physically near to each other that the inductors act as the primary and secondary of a transformer, the tuned circuits are said to be coupled. (Fig. 6.7). When circuits tuned to the same frequency are coupled so that the transformer action is slight (loosely coupled), the selective action of the tuned circuit is increased, and the phase shift of any frequency other than the resonant frequency is large.

At the frequency $f/2Q$ away from resonance, where one tuned circuit is 3dB down, the coupled pair is 6dB down (which is not double the amount, but only $1\frac{1}{2}$ times, when we use 'voltage' dB ratios) and the phase has shifted by 90°. For this reason, coupled tuned circuits are preferred when one signal must be selected from a group on nearby frequencies; the coupled circuit has greater selectivity.

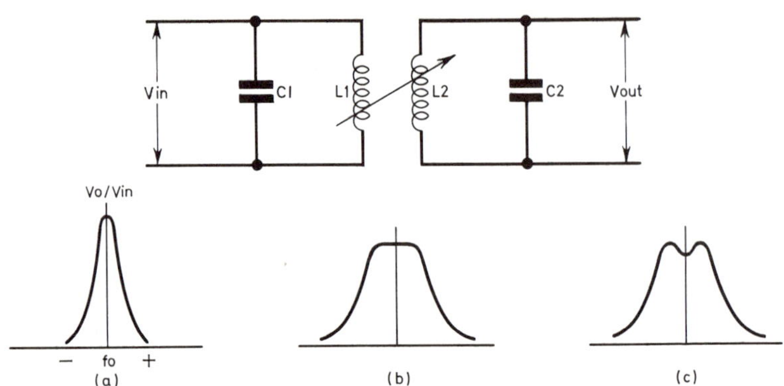

Fig. 6.7 Coupled tuned circuits. (a) loose, (b) tight, (c) overcoupled.

If the inductors are brought closer together, the gain from the primary winding to the secondary output becomes greater but the shape of the curve of gain against frequency changes, becoming flat on top and eventually, when the coils are very close, developing peaks on each side of the resonant frequency. This has happened because the resonance in each coil is shifted by the other, but in opposite directions, and is called *overcoupling*.

For many purposes, the flat-topped curve is useful, since it allows a wider bandwidth to pass, an essential condition in TV and in radar signals if the received pulses are not to be changed in shape, because passing through a narrow-bandwidth circuit has an integrating effect on a pulse. The overcoupled curve can be made flat at the expense of gain by adding resistance in parallel with each tuned circuit. The expression for this was given earlier; it has the effect of lowering Q and so making $f/2Q$ greater. Resistors used in this way are called 'damping resistors'.

Similar effects can be produced in amplifiers using large numbers of tuned circuits by tuning each circuit to a different frequency. This is termed 'stagger-tuning', and is widely used in radar receivers. The amplifier bandwidth in a TV set can be around 6MHz and in a radar receiver it may be as high as 30MHz. The operating frequency (in the middle of the band) must be high enough to permit such bandwidths; generally about 37MHz in the TV receiver and 200MHz or more in the radar receiver.

It is worth noting that the product Gain × Bandwidth depends mainly on the active devices, and is fairly constant for an amplifier no matter what damping or feedback is used. Bandwidth is always obtained at the expense of gain, and vice-versa, unless the design is modified to allow more stages to be used.

Other Effects of Tuned Circuits

So far we have dealt only with the effects of tuned circuits on sinewaves. In some circuits squarewaves or pulses are fed to tuned circuits, and we ought to know what comes out. The effect of the tuned circuit is to select a sinewave out of what is fed to it, and this is precisely what happens. What is more difficult to understand is that the sinewave output may continue for some time after the end of the pulse which has caused it.

In many ways the tuned circuit behaves like the piano strings which sound in response to any loud noise—it is this similarity of behaviour which gives us the word 'tuned'. When a tuned circuit is struck by a pulse or any other rapidly changing waveform, it responds. It can respond only at its tuned frequency, and once current starts to move in an inductor, or charge gathers in the plates of a capacitor, the process tends to continue. It continues in the form of a sinewave which dies away slowly (in the form of an exponential curve) until the amplitude is zero.

A tuned circuit used in this way is called a 'ringing circuit' (because of the similarity to the action of a bell), and the waveform is called a damped oscillation, damping meaning in this case the gradual decrease in amplitude (see Fig. 6.8). The addition of damping resistors makes the decrease more rapid, until a value of resistance is used at which the oscillation is finished after one

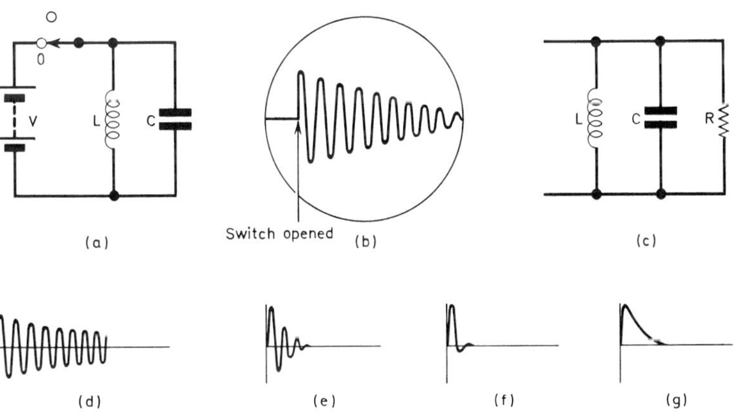

Fig. 6.8 (a) simple ringing circuit, (b) oscilloscope trace showing the action when the switch is opened, (c) ringing circuit damped by resistor R, (d) very little damping, R large, (e) large amount of damping, R small, (f) critically damped, oscillation stops in shortest possible time, (g) overdamped, voltage dies away in time longer than for critically damped oscillation.

cycle or less. This is said to be critically damped; if the oscillation never properly develops, the circuit is 'overdamped'.

Similar oscillations develop in car springs, and the word 'damper' is also used for the device which adds a mechanical form of resistance to the ringing circuit (the spring and the car). The amount of damping is once again measured by the Q of the tuned circuit; the greater the Q value (less resistance) the less damping and the longer the ringing.

Filters

More complex arrangements of inductors and capacitors are known as filters (resistor–capacitor networks are also used, but have very low Q values) because they can be used to filter out signals of different frequencies in the same way that a filter paper can separate liquids from solids.

Four general types of filter are used—low-pass, high-pass, band-pass and band-stop. Low-pass filters (Fig. 6.9) allow all frequencies below a chosen frequency (the cut-off frequency) to pass through the network, the other frequencies above the cut-off being greatly attenuated. The range of frequencies passing the filter is called the pass-band, and the range of frequencies attenuated is the stop-band. Typical low-pass filters use inductors in series with the signal and capacitors in shunt.

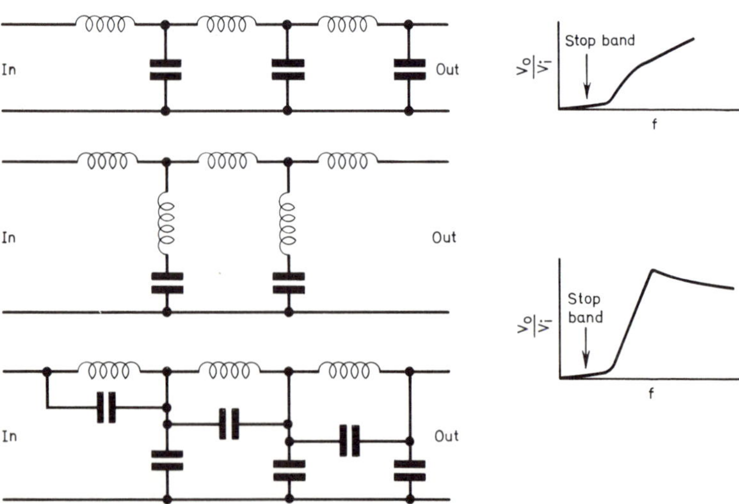

Fig. 6.9 Different types of low-pass filter, and their approximate frequency responses.

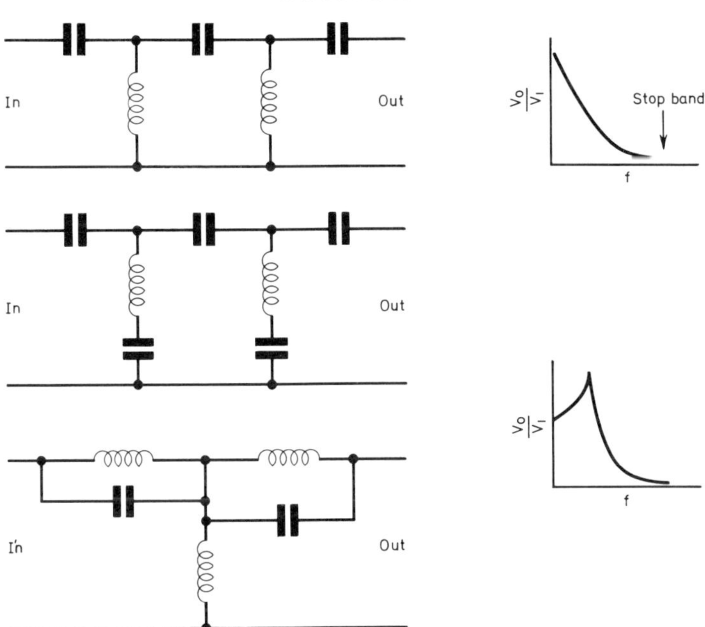

Fig. 6.10 Different types of high-pass filter, and their approximate frequency responses.

High-pass filters (Fig. 6.10) are the opposite of low-pass filters. The pass band is the range of frequencies above the cut-off frequency and the stop band is the range below the cut-off. High-pass filters typically use capacitors in series with the signal and inductors in shunt.

Bandpass filters (Fig. 6.11) pass frequencies between two limits, attenuating all others. They have two stop bands, an upper and a lower. Bandstop filters similarly stop a band of frequencies and have an upper and a lower stop band. Resonant circuits behave as crude bandpass or bandstop filters, but the circuits dignified by the name of bandpass or bandstop behave more accurately than a simple resonant circuit in passing wanted frequencies and rejecting the unwanted.

Filters are specified by their type (low-pass, high-pass etc.), their cut-off frequency or frequencies, frequency of maximum attenuation, where the filter is most efficient at stopping signals, stop band attenuation (which measures the average attenuation in dB in the stop band) and the characteristic impedance.

This last term requires some explanation. In use, filters behave correctly only if the filter ends and begins with the correct impedance,

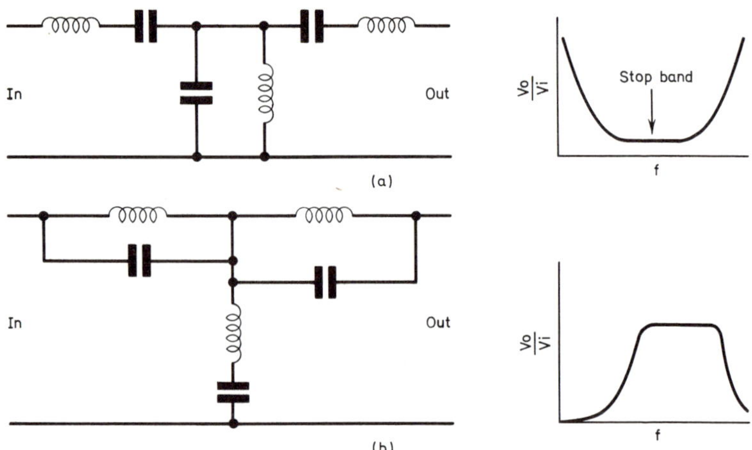

Fig. 6.11 (a) Band-pass filter, one of many possible types, (b) band-stop filter, one of many possible types.

decided by the design of the filter. If this impedance is not provided, the filter is said to be mismatched, and will not behave according to plan, because the signals passed through it tend to bounce back at the terminations and upset the working of the filter. If a filter has a characteristic impedance of 600Ω, there must be an impedance of 600Ω at each end of the filter. This may be an actual resistor or it may consist of the input or output of an amplifier, or even another filter.

When pulse waveforms are passed through a filter, several effects may be noticed:

(a) a low pass filter can act as an integrator or it may delay a pulse so that the pulse arrives at the end of the filter unchanged in shape but some time (1μS or so) later,
(b) a high-pass filter may differentiate the pulse,
(c) a band pass filter usually rings when a pulse is passed through, unless the pass band is very wide or the damping large.

In all cases where pulses are passed through a filter network, some ringing after vertical portions of the pulse must be expected.

Lines

At frequencies of 500MHz and over, it becomes difficult to construct resonant circuits and filters because the values of the inductors and capacitors required are so small. The operations of tuning and filtering can still be carried out using transmission lines, such as parallel lines, coaxial lines or microstrip lines. Such lines or cables can be regarded as long resonant circuits, because the length of wire which constitutes each cable has some inductance, and the inductance is tuned by the capacitance between the two conductors.

The fact that these quantities are 'spread out' along a length of cable, and not lumped into a small inductor or capacitor affects the way in which they behave, however, and the resonant frequency depends on the length of the line rather than on the separate factors of inductance or capacitance which make it up (although these together affect the behaviour of the line) Fig. 6.12 shows various forms of lines.

Lines are tuned by cutting to size and leaving the unconnected end either shorted or open circuit. The resonant frequency of an an airspaced line is given roughly (for a whole-wave line) by: 30/length in cm $\times$ 1,000MHz, if we imagine that the length of the line is occupied by one wave, and the line is shorted at the far end.

If we use half of this length of line, we can still regard it as tuned to the same frequency, because the line will still accommodate a half

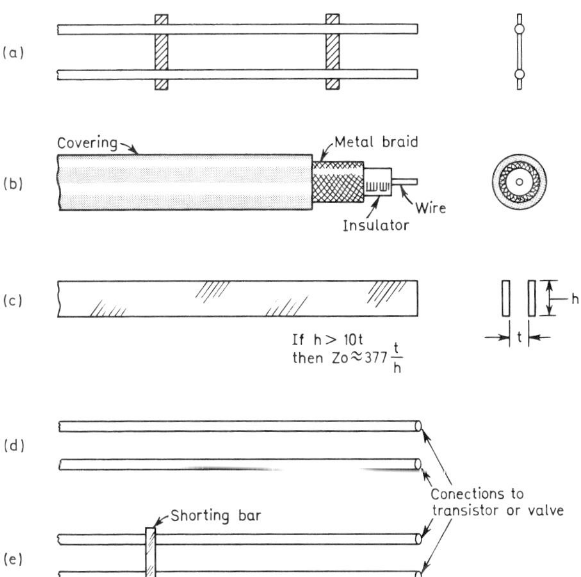

Fig. 6.12 (a) Open parallel lines, impedance about 300Ω, (b) Coaxial cable, impedance about 75Ω, (c) parallel-strip line, impedance about 400 × d/t Ω, (d) open line tungin circuit, (e) shorted line tunging circuit.

wave so that the zero-voltage points are at each end. If we use an open circuit line at one end, the length of wave on the line is only a quarter wave, and the frequency is double that of the half-wave line, given by 7·5/length in cm × 1,000MHz.

Lines cut so as to accommodate a number of quarter or half waves are tuned lines, or resonant lines, and are extensively used in oscillators and in filtering circuits. If lines are terminated by connecting an impedance at each end which is equal to the impedance which the line presents to a wave, then the lines behave as excellent conductors of signal with very low attenuation and a very wide pass-band; the line appears to have no inductance of capacitance, only resistance. The characteristic impedance with which the line should be terminated to ensure this is calculated from the dimensions of the cables which make up the line.

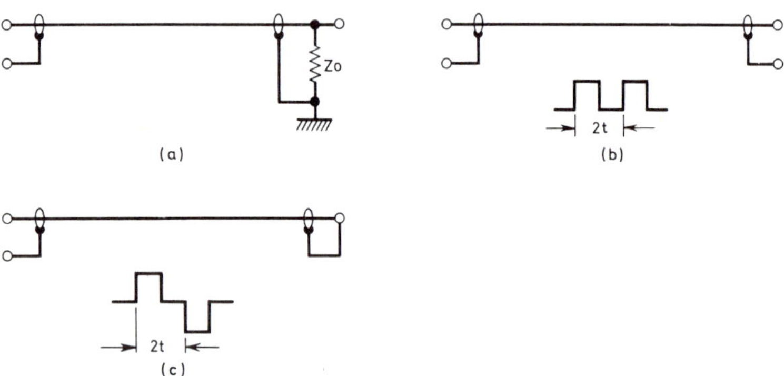

Fig. 6.13 Pulses in lines (a) in terminated line there are no reflections. The line behaves as if it were a resistor of value $Z_0\Omega$; (b) Open circuit line. A pulse sent in at the input is reflected back to the input a time 2t later, where t is the time for a pulse to travel the length of the line in one direction; (c) Short circuit line. The reflected pulse is inverted in this case.

If the impedances are incorrect for any reason, then the line starts to show some resonant line behaviour, waves travel less freely, and high voltages appear at points on the line half wavelengths apart. In this condition, the line is mismatched, and has a voltage standing wave ratio (VSWR) to prove it. When lines are used with pulse signals (Fig. 6.13), open or shorted lines cause pulses to be reflected (after travelling down the line and back) unchanged in shape, but a short circuited line causes the reflected pulse to be inverted. This property of lines is used in generating very short pulses.

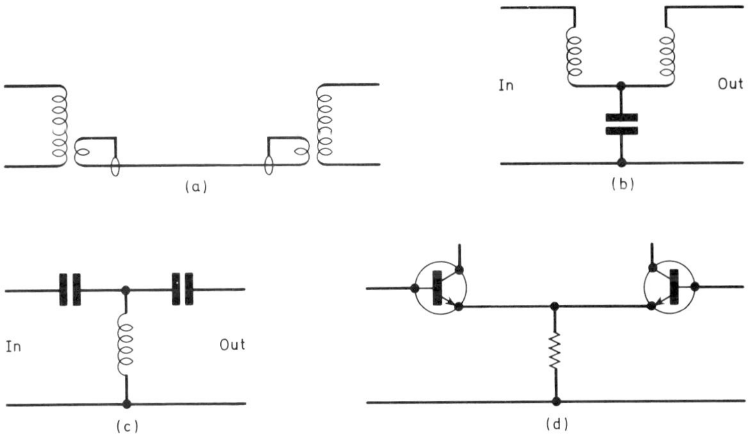

In
Out
(a)
(b)
In
Out
(c)
(d)

Fig. 6.14 (a) Link coupling of two circuits, usually tuned circuits; (b) common capacitor coupling; (c) common inductor coupling; (d) common resistor coupling in the balanced amplifier circuit.

Further Notes

Resonant circuits may be coupled in ways other than the transformer method shown earlier. Link couplings act as 'long distance' transformers so that resonant circuits some distance apart behave as if they were inductively coupled directly. Each end of the link cable is coupled by a low inductance winding which should be equal in impedance to the impedance of the line which is used to form the link.

Common impedance couplings operate by making a resistor, capacitor or inductor (or even a tuned circuit) common to both circuits. Active devices are also coupled in this way in the very common balanced amplifier circuit. Fig. 6.14 shows various couplings.

Piezoelectric crystals can also act as resonant circuits and filters with a very high Q and (consequently) a very narrow bandwidth, sometimes as narrow as a few Hz. A piezoelectric crystal is a material, such as quartz or barium titanate, which changes its dimensions when an electric field is applied to it. A piece of such material can be cut and coated with metal on opposite faces.

An alternating voltage applied between the metal contacts will then cause the crystal to vibrate; conversely vibration applied to the crystal can cause an alternating voltage to appear at the electrodes. Since every object has a mechanical resonance, the vibration is at that frequency. Such devices are used in oscillators, in

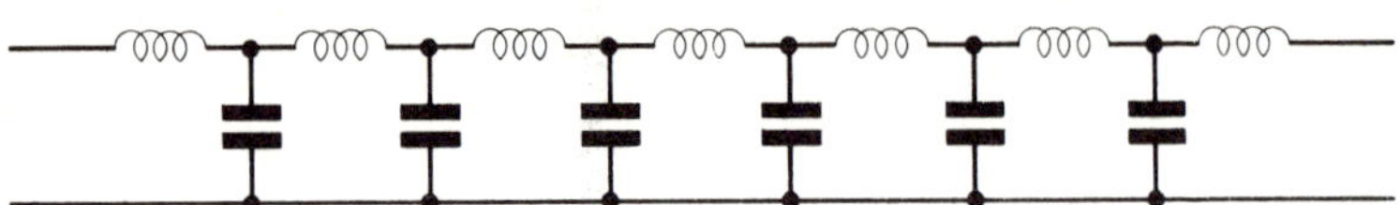

Fig. 6.15 Lumped parameter delay line. The time for a wave to travel down this line is n $\sqrt{LC}$, where n is the number of sections, L and C are the values of each of the inductors and capacitors respectively.

a form of transformer and in delay lines, where the comparatively low speed of mechanical waves is used to cause a time delay in a signal.

A line is sometimes made artificially where an actual length of line would be inconveniently long. Such a line is called a 'lumped-parameter' line (Fig. 6.15), and is made up of series inductors and shunt capacitors. Its characteristic impedance and its equivalent length can be calculated from the values of inductance and capacitance used, and the number of sections.

Inductive 'Kick'

Just as capacitors store up the energy of electrons which are not moving, so inductors store up the energy of electrons which *are* moving. When electrons are moving in an inductor, it is very difficult to stop them suddenly, and when a circuit containing an inductor is open-circuited, the inductor behaves as if electron flow was trying to continue; one end of the inductor becomes drained of electrons and the other end accumulates them.

In ordinary language, there is a voltage across the inductors. The size of the voltage is $L \times$ change of current/time taken. The amount of the voltage can be several kV from a low voltage circuit, and

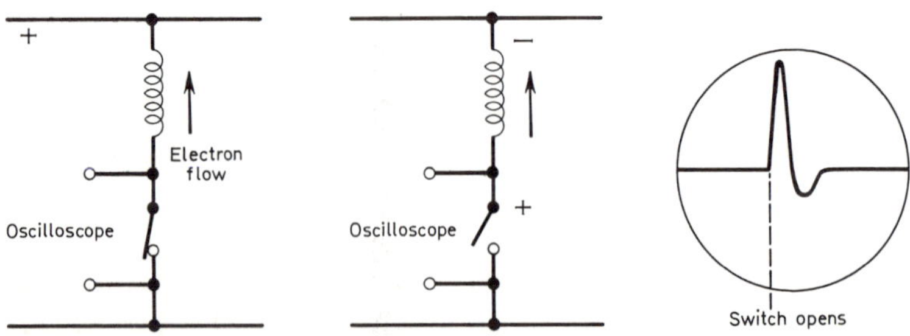

Fig. 6.16 Inductive 'kick'. The circuit behaves as if the electrons did not stop moving when the switch is opened. The oscilloscope connected across the switch records a large positive voltage.

the principle is used to obtain high voltages in car ignition circuits and in TV receivers (Fig. 6.16).

Equivalent Circuit of a Transformer

Using an equivalent circuit of a transformer enables us to predict the effect of using a transformer in a circuit, apart from the obvious effect of stepping voltage up or down. Several equivalent circuits can be devised, some showing the effect on the primary side only, some showing the effect on the secondary only, some showing the effect split between primary and secondary.

The circuit shown in Fig. 6.17 is the equivalent which gives the effect of the transformer and its connections on the secondary side (load) on the primary circuit, so that the secondary load is shown as a transformed load in the primary. The secondary load is said to be reflected into the primary, and other features of the secondary circuit

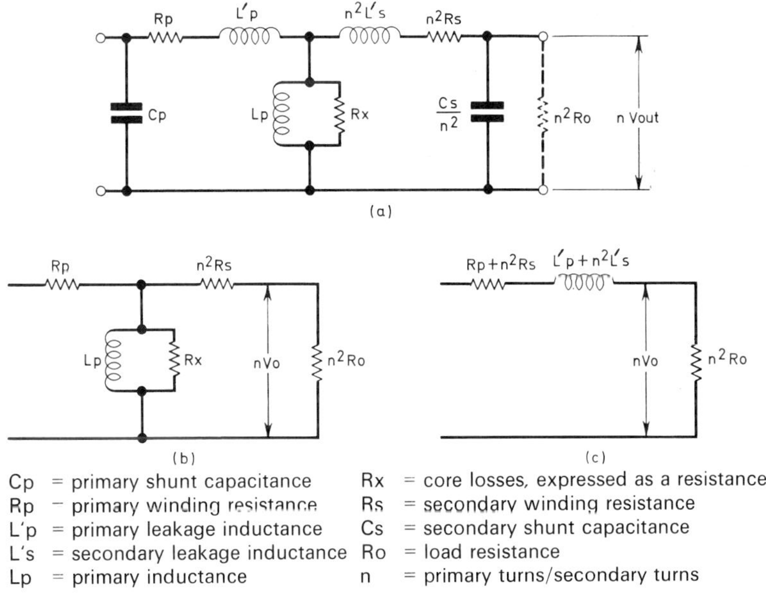

Cp = primary shunt capacitance
Rp = primary winding resistance
L'p = primary leakage inductance
L's = secondary leakage inductance
Lp = primary inductance
Rx = core losses, expressed as a resistance
Rs = secondary winding resistance
Cs = secondary shunt capacitance
Ro = load resistance
n = primary turns/secondary turns

Fig. 6.17 Equivalent circuit of transformer: (a) full circuit if the transformer delivers a voltage V_{out} at the secondary, then the effect on the primary circuit is of the network shown with a signal voltage of $n.V_{out}$ at the terminals. Perfect coupling is assumed; (b) low frequency equivalent; (c) simplest high frequency equivalent.

are also multiplied by the same factor; the factor depending on the turns ratio.

At frequencies which are low compared to design frequency, a simplified equivalent circuit can be used. At frequencies high compared to design frequency, another equivalent can also be used. The low frequency equivalent ignores all inductances except the primary inductance; the high frequency circuit (if we ignore shunt capacitance) is much more affected by the leakage inductance. If shunt capacitance cannot be ignored at the frequency considered, then the complete equivalent circuit must be used.

Magnetic Amplifiers

The characteristics of a material being magnetised, shown in the drawing, can be used as the basis of an amplifier. Normally, for transformer action, we try to use a core material where the characteristic resembles as closely as possible a straight line, with minimum change of slope and smallest possible area of loop. It is possible, however, to make core materials which permit a wide range of characteristics, including some which reach saturation (the point at which the magnetisation due to the core has reached maximum). An inductor wound on such a core is called a 'saturable reactor', and its inductance will vary according to the extent to which the core is saturated (Fig. 6.18).

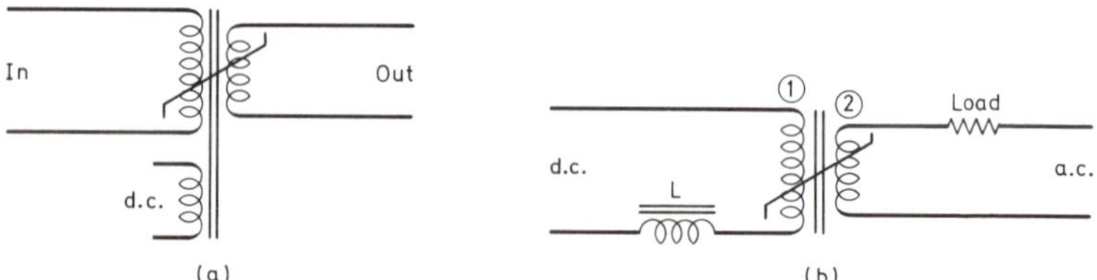

Fig. 6.18 (a) Principle of saturable reactors, (b) control of a.c. in a load by the d.c. in a saturable reactor.

If a transformer is wound on such a core, together with a third winding, then, with no current flowing through the third winding, normal transformer action will take place provided that the core is not saturated by the primary current. If d.c. is now passed in the third winding, the core will come closer to saturation, the primary inductance will be lower and the coupling less so that the output will drop. With the core completely saturated, the output is almost zero.

This can be made the basis of an amplifier circuit which consists of reactors only. The operating power is a.c., and the 'signal' controlling the amplifier is d.c. or very low frequency a.c. Very high gains can be achieved, for a low voltage passing current in the control winding can control a high voltage high current a.c., which if necessary can be rectified and used as the control signal in another amplifier.

Magnetic amplifiers are extremely reliable and are used in industrial control circuits where speed of response is not important, particularly in motor control circuits. They are generally bought as a complete package, sometimes designed to order by the transformer manufacturer, and seem to be almost unknown outside the specialised field in which they have been used.

CHAPTER SEVEN

VALVES

WE HAVE NOTED IN CHAPTER ONE that electric current is due to the movement of charged particles, especially the negatively charged particles called electrons. Towards the end of the last century, physicists began to wonder if these particles could be persuaded to leave the metal wires, in which they seemed to be present in great numbers, by applying heat. Two facts were soon discovered; the electron currents detected in air were very small, and collisions with the large gas molecules cut down the current, and most wires melted before any trace of large electron currents could be detected.

By evacuating (removing air from) the space round the wire, and using wire of the metal tungsten, which melts only at a very high temperature, electrons were finally detected travelling in space outside a wire. This discovery led to the valve, and, because of the greater understanding of materials which also emerged, to the later discovery of the transistor.

Many engineers training today are not likely ever to work on equipment using valves, but those working on equipment designed for high power outputs, very high frequency operation, or industrial use, may find such equipment totally valve operated for some time to come and service engineers are still likely to face the repair of valved equipment in their workshops. For these reasons and because valves illustrate the behaviour of electrons so well, we ought to understand how valves work and how they are used in the circuits where they are still met.

The Diode

Each separate portion of a valve which makes a distinct contribution to the flow of electrons is called an electrode. The simplest valve is the diode, which has two electrodes—the cathode which emits electrons into the surrounding vacuum and the anode which collects these electrons. To emit electrons, the cathode must be hot, red hot at least, and some provision must be made for heating the cathode.

In small valves this consists of a heating coil placed inside the cathode and electrically heated. In large valves the heater and cathode are one, usually thoriated tungsten, and of large diameter by comparison. The heater of each valve must be run at the voltage or current specified to ensure that the cathode is at the correct temperature. It is not considered as an electrode except in the case where heater and cathode are one piece of wire; a type of construction found in large transmitting valves.

Because the electron is a negatively charged particle which is attracted by positive charges and repelled by negative charges, electrons flow from cathode to anode only when the anode is more positive than the cathode. The diode circuit of Fig. 7.1 shows the effect which conduction in one direction has on various waveforms. Only those parts of a waveform which are more positive than the cathode of the diode are allowed to pass through the diode.

A diode used in this way converts a sinewave into a set of positive half-sinewaves which can be integrated to give a d.c. voltage. A diode used to obtain d.c. from an a.c. supply in this way is called a rectifier; practically every valve diode encountered by the service engineer is used as a rectifier.

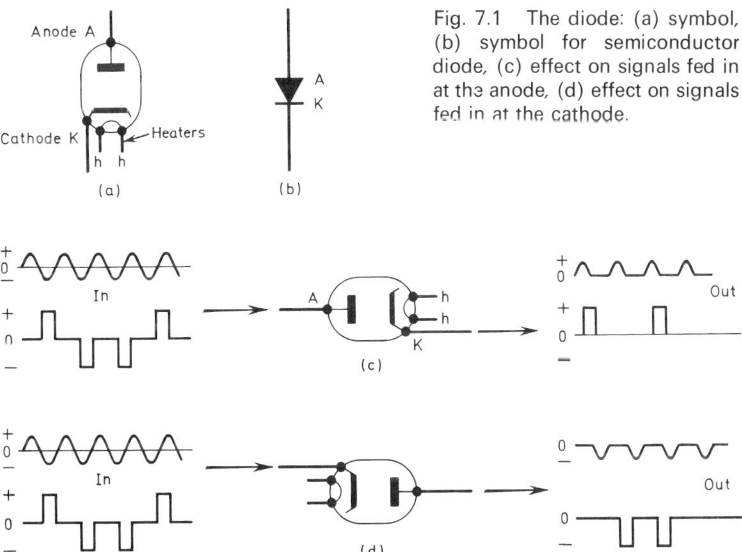

Fig. 7.1 The diode: (a) symbol, (b) symbol for semiconductor diode, (c) effect on signals fed in at the anode, (d) effect on signals fed in at the cathode.

A voltage positive at the anode, and negative at the cathode, causing the diode to conduct, is often called a forward voltage; a voltage positive at the cathode and negative at the anode is called a reverse voltage.

The equivalent circuit of a diode is shown in Fig. 7.2, since the diode behaves as a resistor when passing current, and as an insulator when it is not, there are two equivalent circuits—a resistance of about 500Ω when the diode is conducting and an open circuit when it is not. There is, in addition, some capacitance between anode and cathode, and this is shown also in the equivalent circuit, though it is unimportant unless very high frequencies are being rectified.

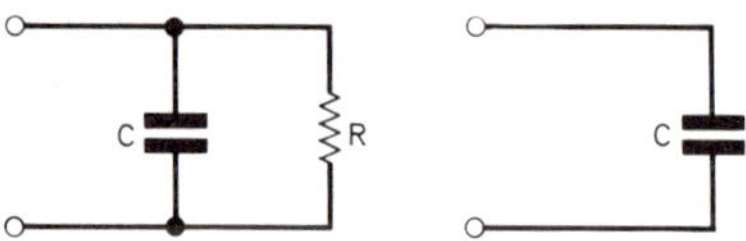

Fig. 7.2 Equivalent circuit of diode (a) while conducting. R is a resistance of around 500Ω, and C is about 15pF. (b) when reverse-biased.

Diode Faults

Valve diodes may suffer from low emission, gas or short circuits. A diode with low emission passes very much less current when conducting than a good diode working at the same voltage. Low emission may be due to low heater voltage (easily checked) or to the cathode nearing the end of its useful life (check by replacement). If the diode cannot be immediately replaced, an extension of life can be obtained by a 10% increase in heater voltage (or current); this is a temporary measure until the valve can be replaced.

Gas in a valve means a leakage of air into the valve interfering with the flow of electrons. Symptoms of gas include a blue glow, visible during operation in the case of a valve with a glass body, very high currents passing for a fairly low forward voltage, and, in the last stages, conduction in the reverse direction as well.

Gas in a rectifier diode usually causes fuses to blow in the mains circuit, due to the high currents flowing when the diode starts to conduct in the reverse direction, and the valve must be replaced. Some rectifiers use a controlled amount of gas to enable them to pass high currents (mercury vapour or Xenon rectifiers). Gas may sometimes be due to using too high a heater voltage, or overheating when the valve is not adequately cooled; the effect is temporary in this case.

Short circuits between anode and cathode are uncommon and

easily detected, since the diode conducts equally well in each direction. The short may exist only when the valve is hot and be open again when the valve is cold. Shorts between heater and cathode can short the valve out if the heater is not fed by an insulated transformer winding. Heater cathode shorts are most serious in equipment using heaters connected in series (TV sets mainly).

Large thermionic diodes used in industrial radio-frequency heaters, X-ray equipment, etc., use thoriated tungsten filaments as an electron source. The wattage required for the filament will be high, requiring currents of up to 150A, and the material is fragile. Particular care must be taken in moving the valve. This fragility is made more serious by the fact that the valve cannot by plugged into a socket like its smaller brethren, but must be securely bolted down, and the leads clamped on.

At the voltage at which these diodes are operated (up to several hundred kV) particular care must be taken over the insulation of filament supplies when the valve is being used to provide a positive supply. As the filament will then have the full d.c. output of the valve, the supply to the filament must be taken from a transformer whose insulation can withstand such a voltage. For this reason, such valves are being replaced with large assemblies of semiconductor diodes in series wherever possible.

The Triode

In a diode, with anode and cathode widely separated, large changes in voltage of the anode cause some changes in the current flowing. If the anode and cathode are close, a very small voltage change in the anode causes large changes in the current flowing. The closer an electrode is placed to the cathode, the greater the effect its voltage has on the current of electrons from the cathode, and the triode valve uses a wire electrode (the grid) close to the cathode to control the flow of current to the anode.

Voltage changes at the anode cause changes in current, but not to such an extent as changes of voltage at the grid. Typically a change of 1 volt at the grid may cause a change of 10mA cathode current; the anode would have to change by 250V to have the same effect. The effects which anode and grid voltages have on cathode current are measured by the figures of Transconductance (Mutual Conductance) (symbol g_m) and Slope Resistance (symbol R_a) for the valve. These figures are not constant, but vary according to operating conditions; they are termed parameters, and the two above are the most important of several possible valve parameters.

Transconductance or mutual conductance is quoted as milliamps of cathode current change per volt of grid signal and in the case quoted the g_m was 10mA/V. The g_m is sometimes quoted in micromhos (μmho) and the figure is 1,000 times that used for mA/V (10mA/V = 10,000 μmho).

The slope resistance is the change in anode voltage divided by the change in cathode current which it causes, and is usually quoted in kΩ, (since $V/I = R$, the g_m is the other way round (I/V) and this is the reason for sometimes quoting the value in upside-down ohms (or mhos) as shown. Fig. 7.3 shows valve characteristics.

Biasing the Triode

Current can flow in the triode only when the anode is positive. The grid, however, is an open coil of wire and, despite its nearness to the cathode, can be made several volts negative before current stops flowing to the anode. In small triodes, the grid is always used at negative voltages to avoid the complication of having

The first letter: Indicates fundamental type of device.	
X	Photosensitive valve or tube
Y	Vacuum valve or tube (not photosensitive)
Z	Gasfilled valve or tube (not photosensitive)
The second letter: Indicates the construction or application.	
A	Diode
C	Trigger tube
D	Triode or double triode
G	Miscellaneous
H	Travelling wave tube
J	Magnetron
K	Klystron
L	Tetrode, pentode, double tetrode, double pentode.
M	Cold-cathode indicator or counter tube
P	Photomultiplier tube or radiation counter tube.
Q	Television camera tube
T	Thyratron
X	Ignitron, image intensifier, image converter
Y	Rectifier
Z	Voltage stabiliser or reference tube.
The reference letters are then followed by four figures which form a serial number. For basic types, the last figure is zero, variations are indicated by the numbers 1 to 9	

Table 7.1—PRO-ELECTRON NUMBERING SYSTEM FOR VALVES (industrial, transmitting, navigation, communications)

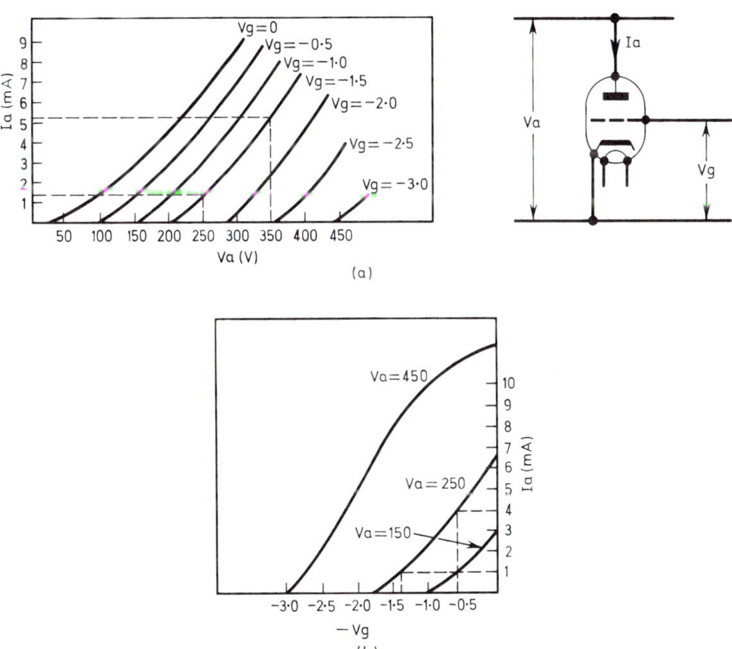

Fig. 7.3 Valve Characteristics. (a) Anode current plotted against anode voltage for values of (fixed) grid voltage. The R_a is found for a particular set of conditions by drawing the lines as shown. For a voltage change from 250 to 350, the current changes from 1·3 to 5·2mA. Under these conditions (−6V grid bias and average anode voltage of 300V), the R_a is 25·5kΩ; (b) Anode current plotted against grid voltage for fixed values of anode voltage. From this graph, g_m can be found at any operating condition. In the example shown, g_m is 3·7mA/V.

current flowing between cathode and grid. In large triodes, particularly those used for transmitters and industrial heaters, the grid voltage ranges from high negative to high positive values.

The values of grid voltage and cathode current used in the valve when no signal is present are known as the bias values. The exact bias values used determine the range of possible input signals, how linear the amplification of the valve shall be, and also the values of g_m and R_a. If the valve has a negative bias voltage on its grid so that no cathode current flows (biased to cutoff), then only positive signal voltages can be amplified.

Conversely, if the grid is connected to cathode or to a positive voltage through a large resistor (otherwise the signal would be shorted) further positive voltage has little effect on cathode current from a low impedance source and only negative signals can be handled. For linear amplification, the bias should be between these extremes, and the input signal smaller than the voltage between cutoff and saturation.

The designer finds the bias conditions which he requires by looking up a set of graphs (called characteristics) which show the effects on cathode current of variations in grid voltage (for several different anode voltages), and of variations of anode voltage (for several different grid voltages). The service engineer is interested only in whether or not these bias conditions still exist, and if the valve is working correctly.

For example, assume an amplifier stage using a small valve is being checked, as shown in Fig. 7.4. The resistor fixed between anode and positive supply (the load resistor) is 15kΩ, the positive supply is +300V and the anode voltage with no signal into the amplifier is +210V, measured with a meter which takes practically no current from the point measured. The voltage across the resistor is 300V − 210V = 90V. and the current by Ohm's Law is $I = V/R$ = 90/15mA = 6mA.

This must be the current passing through the valve. The next step is to look up the valve characteristics which deal with current, grid bias and anode voltage and find the grid bias which permits a current of 6mA to flow at an anode voltage of 210V. The grid should then be biased to this extent, either by having the grid more negative than the cathode or by having the cathode more positive than the grid (which comes to the same thing but is easier to arrange).

If the calculated bias does not agree with the measured bias (note that measurements at the grid may be unreliable because of the current consumed by the meter, see later) then something is wrong. If the bias which exists should be causing much greater current, the valve has low emission. If the existing bias should be causing little or no current, the valve may be gassy or there may be a leakage resistance at the anode. It is worth noting that a given sample of valves may differ as much as 50% from the characteristics printed for that type, and therefore only large differences have much meaning.

The quantity $g_m R_a$ (the transconductance times the slope resistance) is often called the 'Amplification Factor' of a triode (symbol, the Greek letter μ) because this is the maximum voltage amplification which is obtainable from the triode when the load resistance is very high.

Large triodes, used in high power oscillators and transmitters, are usually biased so that comparatively little current flows under no signal conditions, and meters for checking current under working conditions are built into the circuit.

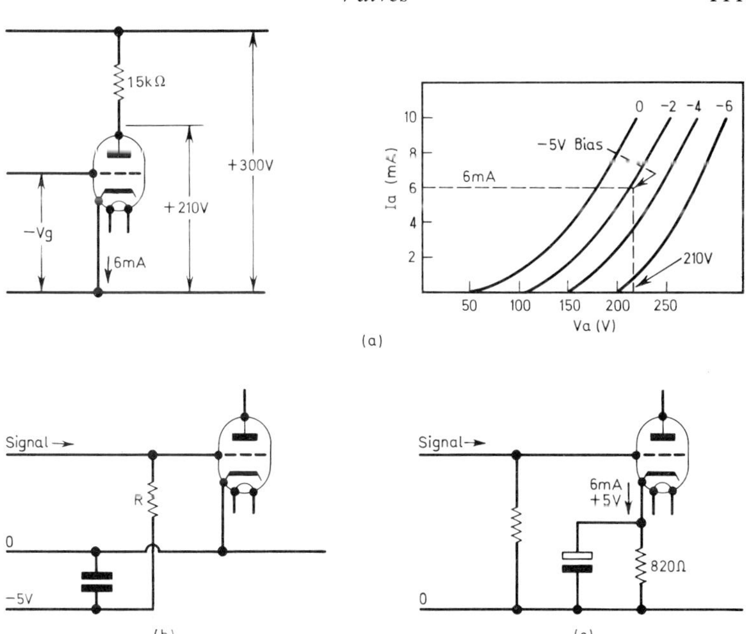

Fig. 7.4 (a) The correct bias (knowing the current passing) can be found from the characteristic; (b) The bias may be supplied through a negative line. R is usually high, 1MΩ or so, and a meter cannot be used to measure the voltage between grid and cathode; (c) if cathode bias is used, the positive voltage at the cathode may be measured with a multimeter. This is equal to the negative bias voltage if the grid is at earth voltage.

The Triode Amplifier

The triode amplifies because a small voltage change at the grid causes a change in the electron current flowing between cathode and anode, and, if a load (resistive or complex) is placed between anode and positive supply, the anode current change is converted into voltage change which can be considerably larger than the change at the grid.

Calculation of amplification is best tackled, as far as the service engineer is concerned, by the equivalent circuit of Fig. 7.5. The input side of the equivalent circuit is of a very high value resistor (thousands of MΩ) with a diode and resistance in parallel. This behaves as the grid circuit of the triode, taking practically no current when signal voltage is applied as long as the grid is negatively biased,

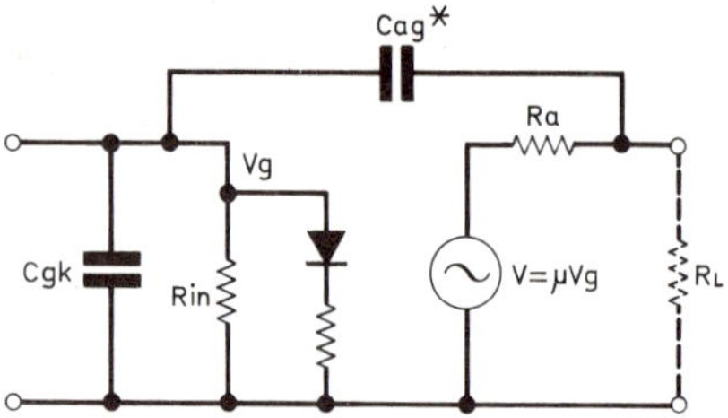

Fig. 7.5 Equivalent circuit of triode. R is very high, several thousand MΩ. Note that V_g and V refer to a.c. *signal* volts and not to d.c. *bias* volts.

C^*_{ag} is the value of capacitance found when the valve is working. If C'_{ag} is the measured capacitance with the valve not in operation, then $C^*_{ag} \times C'_{ag} \times A$, where A is the amplification of the valve in the circuit used. This multiplication of the capacitance is called the 'Miller Effect'.

but behaving as a conducting diode with a small series resistance whenever the grid is positive.

On the output side we have a 'generator' which gives a voltage of μV_g (because the cathode current is controlled by the grid voltage) in series with a resistance R_a which is the slope resistance. The figures of μ and R_a can be obtained for the valve used and the bias conditions in which it is operated by consulting the manufacturer's figures, or by actual measurement on a valve characteristic meter. The equivalent circuit shows many of the features of the triode in a very convenient form:

(1) The input impedance is very high as long as the grid is negative, it is low (about 1kΩ) when the grid is positive to the cathode.

(2) The output impedance is the slope resistance and is usually in the range 5kΩ to 50kΩ (very low impedance triodes exist with R_a about 500Ω).

(3) The load resistor forms a potential divider with the slope resistance so that the actual signal voltage out is, by the potential divider law of Chapter 5,

$$\mu V_g \frac{R_L}{R_a + R_L}$$

Since V_g is the number of volts change at the grid, $\mu R_L/(R_a + R_L)$ must be the number of volts change at the load for one volt change at the grid. It is therefore, the voltage amplification of the valve.

(4) The equivalent circuit can easily be added to, so that other features of the triode can be accounted for. As an example, Fig. 7.5 shows the effects of the capacitances which exist inside the valve.

The grid-to-cathode capacitance (C_{gk}) causes a small time constant at the input but the capacitance between anode and grid (C_{ag}) causes negative feedback which must seriously affect the amplification at high frequencies where the reactance of C_{ag} is small.

This feedback is termed 'Miller Effect' and the capacitance is 'Miller Capacitance'. Diodes at the output can be used to remind the designer that the output voltage is limited by the power supply voltage and the minimum anode voltage at which the valve still works.

Failure of Triodes

In addition to the possible faults listed for diodes, triodes can be put out of action due to bias failure. If the bias voltage becomes too

The type number consists of two or more letters followed by two, three or four figures. The type number gives information on the uses of the valve, the heater or filament rating and the type of base.

The first letter: indicates the heater (filament) voltage or current.

D	0·5V to 1·5V	H	150mA heater
E	6·3V	P	300mA heater
G	5·0V	U	100mA heater

The second and subsequent letters: indicate the general class of the valve.

A	Single diode	H	Hexode or heptode
B	Double diode	K	Heptode or octode
C	Triode	L	Output tetrode or pentode
D	Output triode	M	Electron beam indicator
E	Tetrode	Y	Half wave rectifier
F	Voltage amplifying pentode	Z	Full wave rectifier

These letters may be repeated: CC indicates a double triode, CF a triode-pentode, BF a double-diode-pentode etc.

The first figure: indicates the type of base.

1. Miscellaneous bases
2. B10B base (formerly used for B8G)
3. Octal base
4. B8A base
5. B9D base
6. } Not now used
7. }
8. B9A base
9. B7G base

The remaining figures make up the serial number, indicating the stage of design or development. All valves designed for "entertainment" uses have a serial number of three figures. Valves for other purposes (special quality etc.) have a serial number of four figures.

Table 7.2—PRO-ELECTRON NUMBERING SYSTEM FOR VALVES (valves for domestic receivers and amplifiers)

negative (due possibly to the failure of a resistor in the bias network) the valve may cut off, only positive peaks of signals being amplified if they are above cut-off voltage *or* the valve may be not quite cut off, and negative portions of the signal are clipped.

If the bias voltage becomes positive, or equal to cathode voltage, high current flows between the anode and the cathode. If the load is a high resistance (20kΩ or more), the anode voltage is low, and little damage can be done. If the load is a low resistance (a transformer winding or a low value resistor of, say, 4·7kΩ) both load and valve will overheat due to the high current, and both may be damaged. The current flowing can be calculated if the load resistance is known, since, by Ohm's Law,

$$\text{Current flowing} = \frac{\text{Voltage below supply volts}}{\text{Load resistance}}$$

(Current in mA, Voltage in volts, Resistance in kΩ)

Low emission causes the anode current to be restricted, and amplification is reduced. Gas, in its early stages, causes positive signals at the grid to be amplified more than negative signals, causing severe distortion of a sinewave. Internal shorts of grid to cathode have the same effect as lack of bias; grid to anode shorts are rare and easily detected by the high voltage measured at the grid (unless the grid is connected to a low impedance such as a transformer winding—in this case replacement of the valve is the quickest check).

Table 7.3 is a fault finding chart for valves generally. Note that such effects as o/c or s/c signal paths are not included, since these are not valve faults.

The Tetrode and Pentode Valve

The triode has two important disadvantages as an amplifier. The amplification obtainable is rather low, due to the low value of R_a, and the gain falls off with frequency in a simple amplifier due to the Miller capacitance. In circuits using a tuned circuit as anode load and another supplying the grid, the Miller capacitance causes oscillation instead of amplification. An extra grid called the screen grid between the triode grid (now called 'control' or 'signal' grid) and anode cuts down the Miller capacitance by screening the control grid from the anode, and also causes the slope resistance to be much higher. This type of valve, the tetrode, behaves better than the triode, but oscillates at low anode voltages; it is found now only in the large sizes for industrial use.

The addition of a third grid, the suppressor grid, suppresses the

A valve stage has failed. A signal of the correct amplitude and frequency can be detected by the oscilloscope at the input to the stage, but the signal out is absent or distorted. Check first for obvious disconnections, then check anode voltage using a high resistance meter such as the Avo Mk. VIII. Where an output valve of high power dissipation feeds an inductive load (as in a transmitter) the voltage check must be made with no signal applied. If an oscillator is being checked, the current to the anode should be checked rather than the voltage.

Test Results	**Next Test**	**Normal If —**
No anode volts	Power supply	
Anode voltage is almost equal to positive line voltage	Grid bias (see below). Valve emission. Cathode connection G_2, G_3 of pentode. Anode load value	Input consists of positive-going pulses. Load is inductance, but check for s/c or o/c windings
Anode voltage midway between line and cathode voltages	Output and input connections	
Anode voltage almost equal to cathode voltage	Grid bias (see below). Anode load too high. Leaky output coupling capacitor. Gassy valve.	Input consists of negative-going pulses Input is of small amplitude and low frequency and load is high.

Grid Bias Checks

If the grid is biased by a negative line, this may be at a high impedance, and the voltage should be checked, both on the line and at the grid, using a high resistance valve or MOSFET voltmeter. If cathode bias is used, the voltage between cathode and earth can be checked, using the Avo. If both bias methods are used, the voltage between grid and cathode should be checked directly using a valve or MOSFET voltmeter.

Test Results	**Next Test**	**Normal If —**
Grid volts equal cathode volts	Bias resistor. Bias decoupling capacitor.	
	S/c grid resistor. Leaky coupling capacitor.	Input signals are negative-going pulses.
Grid voltage midway between cathode volts and cutoff.	O/c or s/c in signal path.	
Grid bias at cut-off or beyond.	Stage oscillating. Negative line voltage. Positive connection to cathode.	Input signals are positive-going pulses.

Table 7.3—FAULT-FINDING FOR VALVE CIRCUITS

cause of tetrode oscillation as well as reducing Miller capacitance still further and increasing slope resistance. This type of valve is called the Pentode, and it is used widely for high frequency amplification, but is found only in the smaller sizes due to the difficulty of manufacture in large sizes. The valve shown in 'exploded' form in the photograph of Fig. 7.6 is a pentode used for high gain amplification at audio frequencies (40Hz–20kHz).

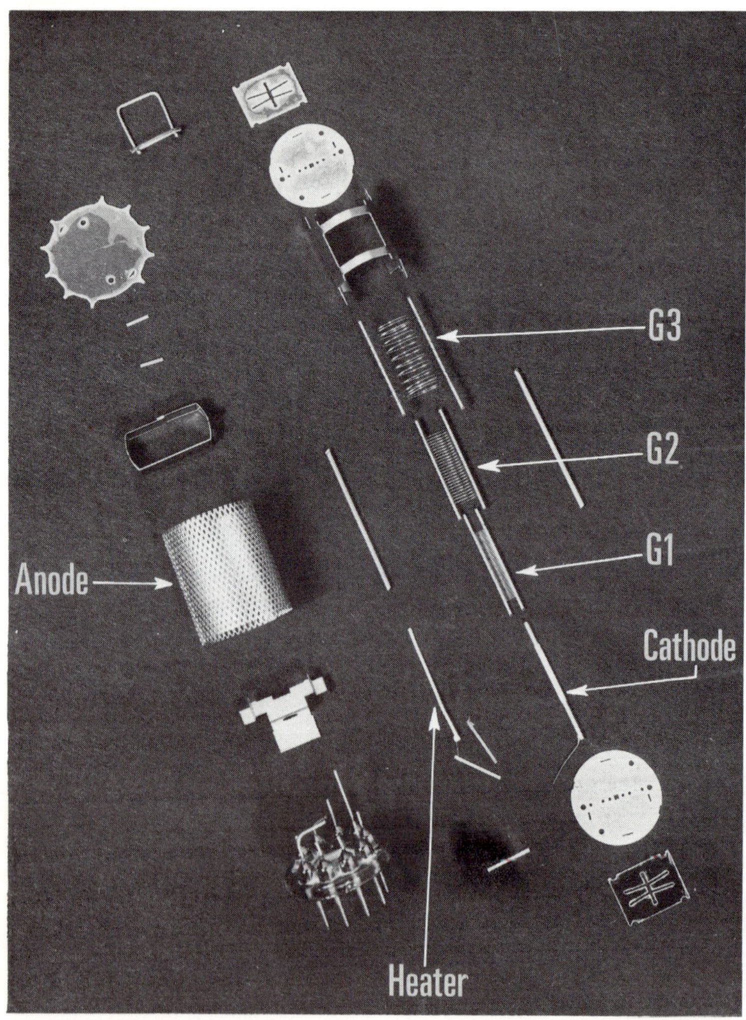

Fig. 7.6 Component parts of a pentode valve, type EF86. (*courtesy of Mullard Ltd.*)

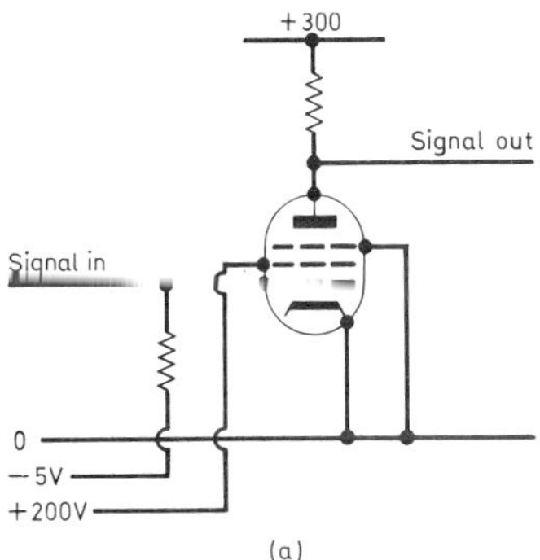

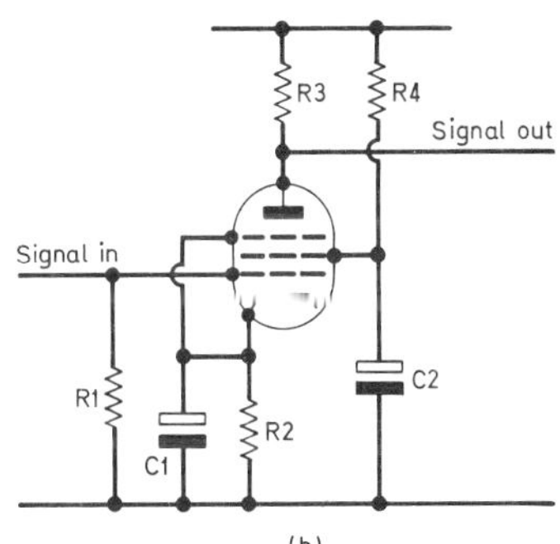

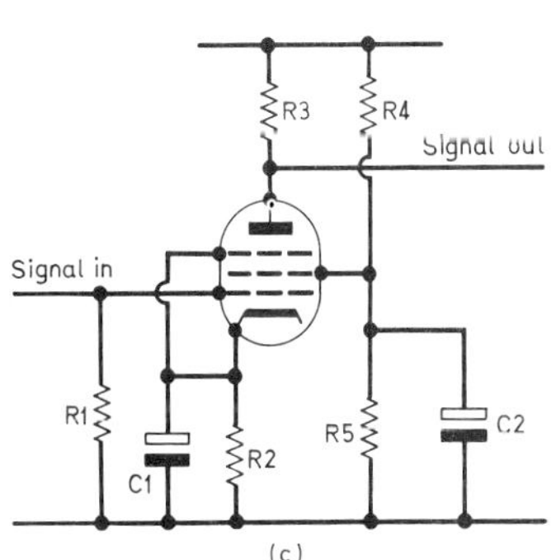

Fig. 7.7 Biasing a pentode. (a) separate bias supplies (seldom seen except on military equipment), useful if a large number of valves are operated at the same bias. (b) Cathode bias, and a dropper for the screen supply. Much used, but the screen voltage is not completely controlled. C_2 prevents loss of gain caused by the signal at the screen reducing the anode signal. (c) Closer control of screen volts. Sometimes R_5 is replaced by a gas stabiliser for complete stabilisation of screen voltage.

Pentode Bias

The suppressor of a pentode is connected to its cathode, unless the valve is used for mixing two signals, in which case one signal is fed in at the control grid, and the other at the suppressor, the bias resistor being returned to cathode voltage. The screen grid is connected to a positive supply which should have a low impedance to the cathode of the pentode.

If, as is common, the screen is supplied from the positive supply through a resistor, a low reactance capacitor is connected between screen and cathode (or earth, if the cathode is earthed directly or by a low reactance capacitor). If this capacitor fails, the gain of the valve drops considerably because the amplified signal voltage at the screen causes negative feedback to the control grid. Fig. 7.7 shows a variety of bias methods which may be found in use.

The Pentode Amplifier

The pentode does not behave like a triode: the high slope resistance means that the control grid voltage has a very much greater effect on current than the anode voltage. Though screen and suppressor voltages also affect valve current considerably, these are usually fixed in the case of straightforward amplifiers. It should be noted that cathode electron current flows to the anode and to the screen, and the currents measured at anode and screen depend on the screen voltage, rather than on the anode voltage. This fact is used in some pulse circuits. When cathode biasing is used, the cathode current is the sum of anode and screen bias currents.

The equivalent circuit of the pentode is shown in Fig. 7.8. The input is as for a triode, but the output is a generator of current $g_m V_g$ (not of voltage, as it was for a triode), and the slope resistance does not appear. The higher the load resistance used, the greater the amplification, though it must be remembered that very high value resistors mean very low bias currents, unless the positive supply is of a very high voltage. As for the triode, the equivalent circuit can be modified to include capacitance between grid and cathode and between anode and grid (though this is very small, about 0·2pF).

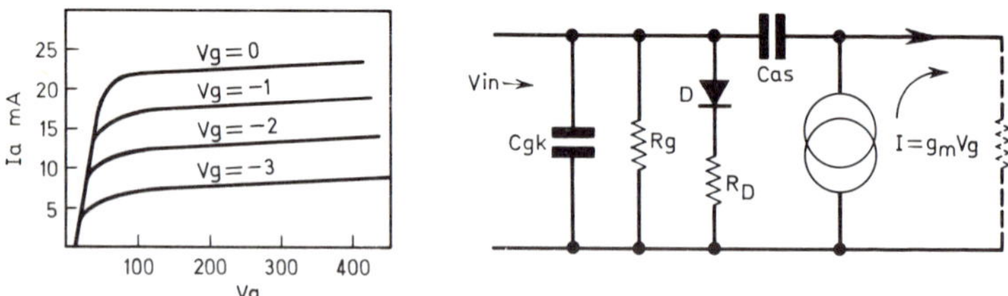

Fig. 7.8 Equivalent circuit and Va/Ia characteristics of a pentode. The same equivalent as for a triode *can* be used, but the very high output impedance of the pentode means that change of anode voltage has very little effect on the anode current. For this reason, the equivalent is of a *current* generator controlled by the grid voltage.

Pentode Faults

Besides those faults noted for triodes, a pentode may be faulty due to failure of screen bias, or bias on suppressor, as well as short circuits between anode and suppressor, suppressor and screen, screen and grid or grid and cathode. Pentode failure is also covered in Table 7.3.

Fig. 7.9 A high-power triode for transmitter use. Filament (heater) voltage is 27·5V and current is 600A. Maximum cathode current is 100A, maximum anode voltage is 17·5kV. Amplification factor is 55. The valve is air cooled, using a fan to force air past the fins on the anode.
(*Photo courtesy of S.T. & C. Ltd.*)

Choice of Pentode or Triode

Several factors can influence a designer's choice of valve. The noise generated by a valve may be important in the case of the first valve in a high gain amplifier, since the noise of the first valve is amplified by all the others. When the signal is fed into or taken from the cathode (as is frequently found for high frequency or pulse operation) the difficulty of keeping screen and suppressor voltages fixed relative to cathode voltage makes the use of triodes preferable. For many industrial purposes, the pentode circuit using a resistor and a capacitor to feed the screen means two more components to go wrong, and a triode is preferred. (Fig. 7.9).

In the larger sizes, the mechanical difficulties of constructing and supporting the extra electrodes influence valve design in favour of triodes again. Most of the higher power circuits are oscillators, where the use of triodes is preferable anyway, and where r.f. amplification has to be used, there are well-known techniques for overcoming the triode disadvantages.

Multigrid and Multiple Valves

Valves having 4, 5 and 6 grids (hexodes, heptodes and octodes) have been made and used, but are seldom found in electronics work (as distinct from domestic radio). Where such a valve is found, the purpose is practically always to mix two signals together so that one modulates the other and is added to the other, or to gate one signal by another, as indicated in Fig. 7.10. In such cases, signals are applied to two of the grids and the others are used as screens.

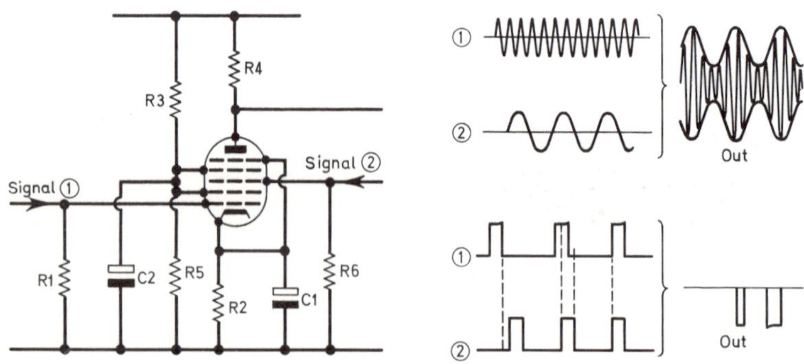

Fig. 7.10 Outline of heptode circuit. Sinewaves fed in as signals (1) and (2) will appear mixed at the anode. If grids (1) and (3) are biased to cut-off, and positive pulses are used as the signals in, then there will be an output from the anode only when the input pulses overlap in time.

Multiple valves consist of two or more separate valve assemblies in one glass body, often sharing a common heater. The number and types of valves which can be mounted together in this way is limited only by the number of pins available on the base, and several special bases were produced for a number of American valves each of which combined three of four types in one envelope (e.g. the *Compactron*, trade mark of G.E. (U.S.A.)). Multiple valves assist in making circuit construction more compact, and are therefore favoured by designers.

Tables 7.1 and 7.2 (pages 108 and 113) show the Continental Valve Coding system which has the great merit of using the valve code numbers to give some indication of the valve type.

Further Notes

The name 'beam tetrode' is sometimes encountered for a type of power valve which is really a pentode which uses plates to guide the electrons to the anode instead of a third grid. In this case, as with several pentode types, the suppressor is connected to the cathode within the valve.

It is common practice now to refer to the grids of a valve by numbering them in order of distance from the cathode; control grid is G1; screen G2; suppressor G3; and this system is also used on electron beam tubes such as cathode ray tubes.

Though electrons from the cathode travel to the anode, cathode current is not always equal to anode current. In a pentode, the anode current is less than the cathode current because of the screen current (cathode current = anode current + screen current). Also in both pentodes and triodes, if the control grid is allowed to become positive and pass current, this again is supplied by the cathode at the expense of anode and any other electrodes.

Where the screen of a pentode is not decoupled by a capacitor, the screen behaves as the anode of a triode, and a signal at the control grid appears in amplified form at the screen. This 'triode' has a g_m and R_a and the product is called the 'inner μ'. Designers sometimes use this 'inner triode' effect to split a signal two ways in one valve.

Some valves may be met which work with very low anode voltages (12V–20V), and no bias voltage. They were used in car radios and in mixed transistor/valve circuits before being replaced completely by later types of transistors.

The most recent series of valves made in the receiving-valve class have used the techniques which have for a long time been associated with the larger sizes. Known as 'frame–grid' valves, they use a rigid framework to ensure that the grid wires are kept at the correct separation from the cathode, so enabling much closer cathode-to-grid spacings to be achieved. This, in turn, gives much higher g_m values, thus enabling the designer to use fewer valves for a given degree of amplification.

Frame–grid valves may be encountered in older black and white TV receivers and in military equipment of the just pre-transistor era, but are unlikely now to be encountered in new equipment except for some very specialised tasks. They are more likely to fail through internal short circuits than the previous generation of valves.

Miniature valves in ceramic or glass envelopes and with unorthodox pin arrangements may also be found in some early u.h.f. equipment, and larger versions will still be found in u.h.f. transmitters. They are mainly triodes.

Klystron and Magnetron

Valve oscillators and amplifiers of conventional design cannot cope with frequencies much above 1GHz (1,000MHz) because the time taken for an electron to travel from cathode to anode is so long (one nanosecond or so !) compared with the time of a cycle. For such frequencies, a different method of amplification is needed, and this is provided by the microwave electron beam tubes such as the Klystron, the Magnetron and the Travelling Wave Tube.

The klystron and the magnetron work on a principle rather similar to the production of a musical note by blowing across the top of a bottle (flute players should understand the klystron easily!). In the klystron, a beam of electrons is passed across the mouth of an accurately machined cavity in a metal block. The cavity acts as a tuned circuit of very high frequency and very high Q, and is affected by the electron beam.

If a signal is injected into the cavity by a loop of wire led inside, that signal will modulate the electron beam making the electrons bunch together in groups, each group representing a peak voltage of signal at the cavity. The electrons continue to bunch up after passing the cavity, and, if they are made to pass over another cavity when they have bunched into as tight groups as possible (and before they scatter again), then more signal can be tapped from the second cavity (by a wire loop) than was injected into the first.

This is an amplifier klystron. If the cavities are coupled together so that the signal is fed back in phase, the klystron acts as an oscillator and the most important type of klystron oscillator is the reflex klystron, in which the beam is reflected at a negative electrode so as to pass the same single cavity again.

The klystron is tuned by altering the dimensions of the cavity or cavities by making one portion flexible or by altering the electrical behaviour by screwing a metal stud further into the cavity. Since the whole tube, including the cavity, must be evacuated, this involves some mechanical complications.

The magnetron is a high power oscillator, which also uses an electron beam, this time travelling in a circular path in the field of a powerful magnet, to excite oscillation in a set of cavities. Very high peak pulse powers can be obtained using a magnetron (5MW pulses with some types), and very high pulse voltages are applied to the cathode, which is pulsed negative. The anode is left at earth potential so that it may be connected to the waveguide which leads the signal out.

The Travelling Wave Tube

This is a low noise microwave amplifier which uses a straight focused electron beam; and once again, tries to modulate the electron beam with a microwave signal. Unlike the klystron, which is sharply tuned (high Q cavities) to one frequency, the TWT is a wideband amplifier, which operates on the principle that an electron beam and a microwave signal will affect each other if they travel together for some time at the same speed.

As the microwave signal travels at the speed of light in space, rather less in wires, but still some ten times faster than an electron beam accelerated by only a few hundred volts, something must be done to slow down the microwave signal so that the conditions for interaction can be met. This is done by making the microwave signal travel along a spiral delay line and shooting the electron beam along the same axis around the delay line. The signal modulates the electron beam, and appears amplified at the anode or collector.

It is essential for the working of the TWT that the electron beam should be 'in focus' (as close to the spiral delay line as possible) along the length of the tube, and magnets (permanent or electro-magnets) are used to ensure this. Many tubes are sold complete with magnets in one package which must not be separated.

The BWO (Backward Wave Oscillator) or Carcinotron resembles, in principle (though not in shape) the TWT and is an oscillator in which the electron stream and the 'signal' move in opposite directions.

Failure of Microwave Tubes

As with any vacuum device having a hot cathode, the emission of the cathode can fail, and in this respect klystrons and TWTs are similar to any other beam device. Magnetrons, however, are unusual in that the heater is switched on only when warming up; once the magnetron is working, so much power is dissipated that the cathode remains hot without any heater supply. Magnetron failure is usually abrupt and complete.

In TWTs the end of life may be noted by a rise in the noise generated by the tube. The magnets of both TWTs and magnetrons must be treated with great respect and no magnetic material placed near them, since a change in the magnetic field can affect the performance of the tube. As with all microwave work, the circuit, whether waveguide or coaxial, should be checked before suspecting the valve concerned.

A vital measurement in microwave circuits is the 'standing-wave ratio' which indicates how much of the signal fed in at a point travels to the other end of the waveguide and how much is reflected back. If too much power is reflected back, the operation of any microwave tube will be seriously upset, sometimes to the point of causing failure.

Photocells and Photomultipliers

Photocells are used for converting light signals into electrical signals. Unlike the other vacuum tubes described so far, they do not obtain their electrons from a hot cathode, but from a material which releases electrons when struck by light; a photocathode. These electrons may be gathered by an anode, in a simple photocell, or they may pass into an electron multiplier in a photomultiplier.

The straightforward photocell is not very sensitive to light, and the electrical signal requires a very great amount of amplification to be of any use. An ordinary amplifier (valve or transistor) introduces a lot of noise into a signal. Fortunately there is a method by which an electron beam signal can be amplified greatly (as much as one million times) with very little noise added; this is multiplication (Fig. 7.11).

The electron beam multiplication can be carried out by the effect of secondary emission, where electrons striking a surface dislodge other electrons. If the voltage between the cathode and the surface struck by electrons is anywhere between about 70V and 4kV (depending on the substance to some extent) more electrons leave than arrive. For most substances the gain is small, 1·1–1·3 or so, but for several metals (silver, magnesium, nickel) coated with the metal caesium the gain in electrons can be as high as 6–10 for each surface struck.

In a photomultiplier, each electron from the photocathode is

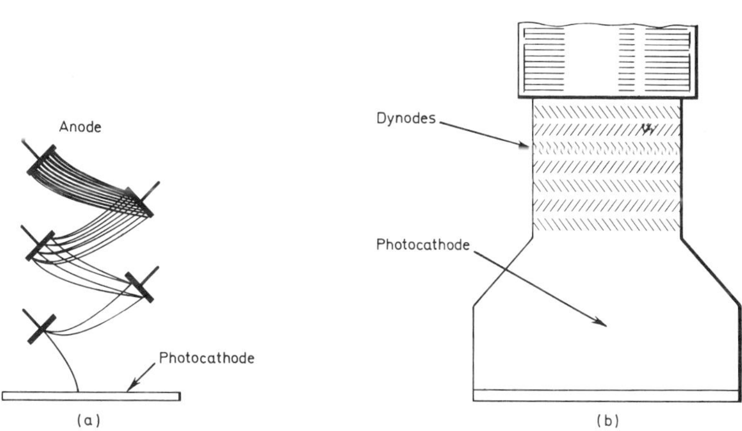

Fig. 7.11 Photomultiplier. (a) action of dynodes, (b) arrangement of dynodes in a photomultiplier, below—actual device (*Photo courtesy of 20th Century Electronics Ltd.*)

accelerated to the next dynode. If each electron causes 5 to leave the second dynode, then 25 electrons are accelerated to the next dynode and so on. In this way, a photomultiplier with a few dynodes can have a very high gain and yet very low noise.

Photocell and Multiplier Circuits

Shown is the type of circuitry associated with photocells and photomultipliers. Since the currents flowing in the tubes are very small, the load resistors must be high to ensure a reasonable voltage, yet if the light input contains high frequency signals the load resistor must be low to avoid the frequency response dropping off too rapidly at high frequencies. The choice of load resistor is always a compromise, especially on photocells intended for high frequency signals, but the amplifier must have a high input impedance and may also need to be corrected for frequency distortion caused by a high value of load resistor.

When a photomultiplier is used, each dynode must be held at a high voltage above the one before, and the circuit must be capable of feeding current to the dynode (which is losing electrons) so that the d.c. voltage is held steady on each dynode. As any signal at the photocathode appears in amplified form at each dynode, some precautions must be taken to prevent signal feedback from the high-voltage dynodes to others lower in the chain, and for this reason, each dynode supply resistor is decoupled by a capacitor to earth (or other common point).

Failure of Photocells and Multipliers

With advancing age, the sensitivity of the photocathode and the dynodes becomes less, so that the tube gradually becomes too insensitive for further use. A short extention of life may be obtained by raising the voltages on the dynodes (so long as the voltages are within the limits specified by the maker of the tube).

Abrupt failure may be caused by internal flash-overs causing the dynode surfaces to be evaporated away in places and so cutting

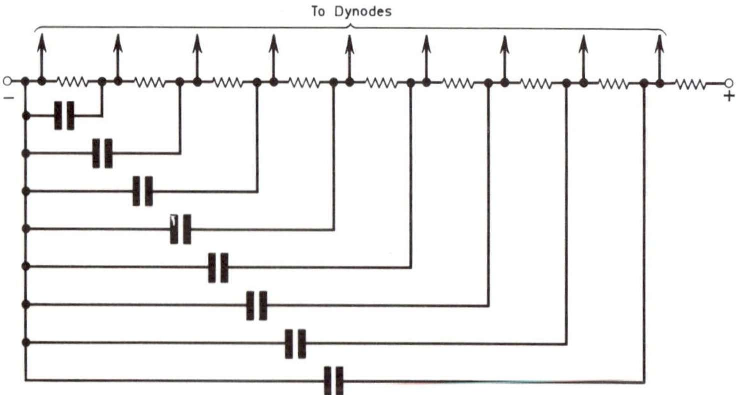

Fig. 7.12 Voltage supplies to dynodes.

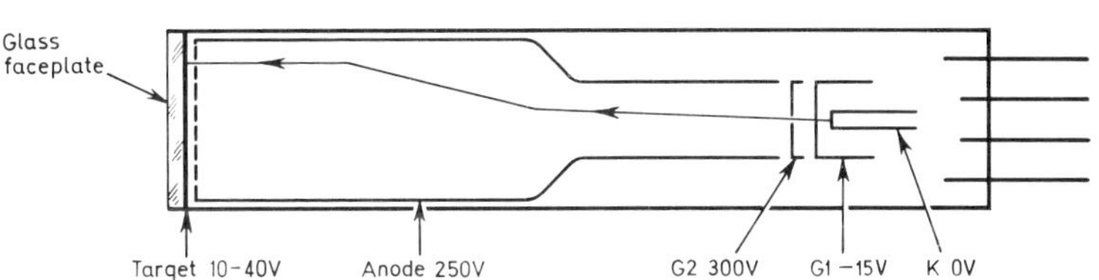

Fig. 7.13 The Vidicon and its approximate voltages.

down the sensitivity very greatly. Most circuits limit the flashover current so much that damage is unlikely, and the usual cause of complete failure is mechanical damage.

The Vidicon

This is a television camera tube used extensively in industrial TV systems. It is shown in outline (Fig. 7.13) with its operating voltages. The target layer deposited on the glass faceplate is photoconductive, meaning that its electrical resistance becomes lower when light falls on it. The glass on which it is deposited is made to be electrically conducting, and is connected to a 40V supply through a load resistor. The other side of the photoconducting target is struck by the electron beam from the gun, which is focused magnetically and electrostatically and magnetically scanned.

When the electron beam lands on the target, the electrons remain without knocking others out because of the low voltage, 40V above the voltage of the cathode (contrast this with the secondary emission in a photomultiplier), and the voltage on the gunside of the target is lowered to zero. Under these conditions no more electrons can land, since there is no longer a voltage between target and cathode, and all electrons are collected by the mesh.

When a light pattern (for example, of dark and light bars) is projected onto the target, the brightly lit portions of target will be at 40V between scans, because of target conductivity, and the dimly lit at a low voltage (perhaps 5V) because the conductivity is low in dimly lit areas. When the beam scans again, a large current (electrons) will flow to the target in the 'bright' areas and a small amount in the 'dim' areas (this current flows also in the target load resistor). From this load resistor, therefore, a video signal can be taken to a suitable amplifier.

The vidicon is a small ($1\frac{1}{8}$in. diameter × $6\frac{1}{4}$in. long) tube with good resolution (meaning that closely spaced dark and light bars can be readily distinguished, instead of appearing as a grey blur), long life, and requires fairly simple circuitry. Its main defect is that moving images in poorly lit scenes appear to leave streaks behind in the same way that a moving object appears blurred on a snapshot. This matters little in industrial TV applications. Apart from mechanical damage or exposure to grossly excessive lighting (such as a flashbulb), which might damage the target, the vidicon fails only by loss of cathode emission.

The Image Orthicon

Probably the most complex type of beam tube made, apart from its near relative, the Isocon, the Image Orthicon is met only in high quality TV cameras designed for studio work or for some medical purposes, as for example, image amplifiers or intensifiers used in conjunction with X-ray equipment. The tube is shown diagramatically, in Fig. 7.14.

The light image on the photocathode causes electrons to be emitted and to strike one side of the glass target, 0·0001 in. thick. By secondary emission, using a metal mesh 0·001in. away from the target as collector, the target is charged most positive in places corresponding to the most illuminated portions of the photocathode. The target is very thin and is made of a conducting glass (conducting by electron flow in modern examples) so that the voltage 'image' appears on the other (gun) side.

Here the beam from the gun approaches the target at a very low speed (because the voltage of the target is only about a volt more than that of the cathode) and lands on the positive areas, being reflected, like light in a mirror; from the more negative areas. The reflected beam passes back down the tube and strikes a series of multipliers, exactly similar to those used in a photomultiplier, emerging finally as a signal at the anode.

The image orthicon has many disadvantages. It is bulky (maximum diameter 4½in., length 20in. for the larger types, 3in. and 16in. for the smaller), is fragile, is very expensive to make and test and has a short life (though this has improved of late). It also suffers from the disadvantage that maximum beam returns on the regions corresponding to minimum light, so that the tube noise is greatest in the poorly lit areas where it is most obvious.

Nevertheless, for overall quality of output signal the image orthicon has never until recently been approachable, and it can be used over a very large range of lighting conditions. Failure can occur in the photocathode, target, either of the two meshes, the multiplier section or the thermionic cathode; even holding the

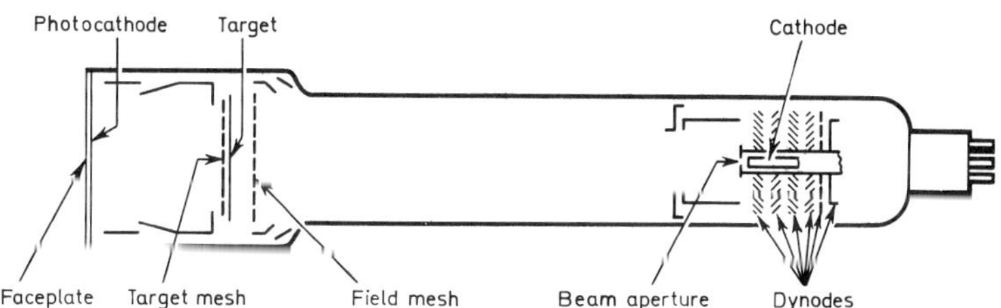

Fig. 7.14 Basic construction of the Image Orthicon.

tube face down can cause irreparable damage if a portion of cathode should fall and pierce the glass target.

The Isocon is a more complex form of image orthicon in which the return beam is split by a deflector plate into two portions, one of which is steady, the other carrying the signal. It has greatly superior signal to noise ratio compared with the image orthicon.

The Plumbicon (trade mark of Philips Ltd.) is a vidicon which uses a different material (lead oxide) for its target. Many of the disadvantages of the vidicon have been overcome, and plumbicons appear to combine a picture quality approaching that of the image orthicon with the simplicity of handling of the vidicon. Colour cameras, which require a minimum of three camera tubes, use plumbicons extensively.

CHAPTER EIGHT

ELECTRONIC DISPLAYS

WHEN A STREAM OF ELECTRONS travels in a vacuum, it can be deflected by magnets or by plates connected to different voltages, and it can cause changes within materials struck by the electron stream. Some of these changes are visible. Some substances glow when struck by electrons, some change colour permanently; other changes are invisible but just as important, such as secondary emission, where a substance struck by electrons breaks up to give off more electrons.

Before the electron was known, all these effects were said to be due to 'cathode rays' which travel from the cathode to the anode of a gas discharge tube. The idea of cathode rays has been left behind, but the name is still used in the term 'cathode-ray tube' (c.r.t.).

Cathode Ray Tubes

A cathode ray tube consists of an electron gun, which produces a stream of electrons of small diameter (0·005in. or less), a focusing deflection stage which focuses the stream (or beam) of electrons on to a screen, and deflects the beam to any part of the screen, and the screen which glows when struck by electrons. In c.r.t's used for television reception or monitoring, the tube consists only of the gun and the screen and the deflection is carried out by coils (acting as electromagnets) placed outside the tube.

The focusing may be by an external coil (electromagnetic focus), a permanent ring-shaped magnet (permanent magnet focus) or by the application of a voltage to an extra electrode in the gun (electrostatic focus). Magnetic deflection has the advantage that less effort is needed to deflect the electron beam from one side of the tube to the other, though the speed of deflection cannot easily be changed over a wide range.

For oscilloscope use, a wide range of deflection speeds is essential, and the effort needed is small, because the tubes used have smaller screens (3–7in.). For this reason, electrostatic deflection is used, and the deflector plates are built in as one unit to the electron gun, or built in to the neck of the c.r.t.

Deflecting and Focusing

Electrostatic deflection uses one pair of plates for each direction in which the beam must be deflected, as shown in Fig. 8.1. When one plate is positive and the other negative, the electron beam moves in the direction of the positive plate. One plate may be held at a constant voltage and the other varied, but this causes a distortion of the spot seen on the screen (called astigmatism) so that its shape is circular only at one part of the screen and elliptical at other places.

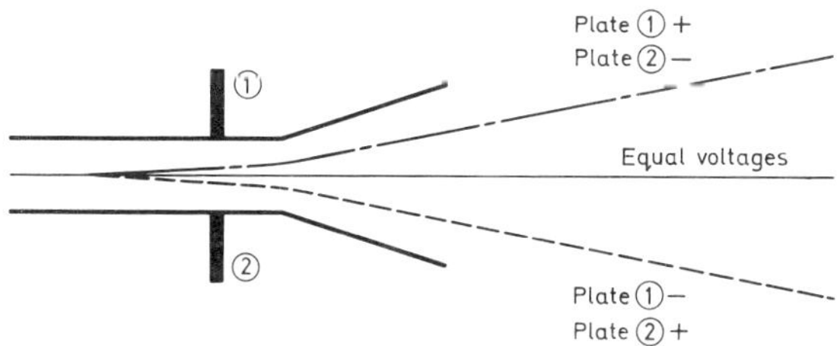

Fig. 8.1 Effect of electrostatic deflecting plates.

When two sets of deflection plates are set at right angles, the spot can be moved to any position on the screen. By convention, we refer to X-plates and Y-plates so that the tubeface is thought of as a graph with an X-axis and a Y-axis. For oscilloscope purposes, the waveform being displayed is applied to the Y-plates and they are made so that a smaller voltage difference is required for deflection (they are closer to the cathode of the electron gun, and sometimes longer than the X-plates.

The sensitivity of deflection plates may be quoted in mm/V meaning that one volt between the plates will cause a deflection of N mm (25·4mm = 1in.). A deflection sensitivity of 0·5mm/V means that 1V deflects 0·5mm; and that 152V is required to deflect the electron beam over the whole of a 3in. (76 mm) tube face.

This sensitivity depends on the accelerating voltage between the final anode (which may be the screen connection or the carbon coating within the tube) and the tube cathode. The higher this voltage, the lower the sensitivity, so that it is also common to see sensitivities quoted as V/mm/V_a, with V_a in kV. (For example: 4V/mm/V_a means that with 1kV on A, a deflection of 1mm requires

4V, 1cm requires 40V (1in. = 2·5cm, requiring 100V) and for 2kV on V_a these voltages would have to be doubled.

Magnetic deflection requires scanning coils, and the direction of the current in the coils (in the traditional sense of current flowing + to −) is the direction of deflection (see Fig. 8.2). The sensitivity depends on the square root of the accelerating voltage, so that four times the accelerating voltage only halves the sensitivity (under such conditions, an electrostatically deflected tube would have $\frac{1}{4}$ of its sensitivity) so that magnetic deflection is favoured when high accelerating voltages must be used.

Magnetic deflection brings circuit problems, however, because the current in the scanning coils must be changed rapidly in the course of scanning, and this causes high voltages (see 'Inductive Kickback') to appear across the coils.

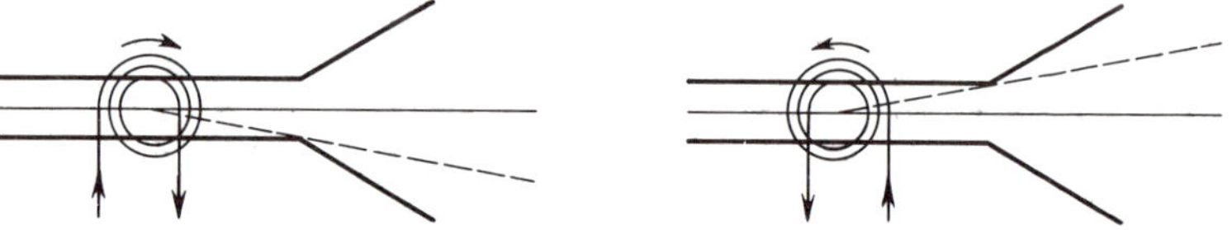

Fig. 8.2 Effect of direction of current in winding on direction of deflection. Note that coils are seldom of this simple 'pancake' shape—the deflecting portions are usually straight, with the joining sections curved to fit closely to the tube.

Post Deflection Acceleration

Since the brightness of the trace of a c.r.t. depends on the accelerating voltage, but decreases as the speed of the trace is increased, designers of c.r.t's intended to display fast traces must use high accelerating voltages so that the traces are visible. Simply increasing the accelerating voltage, however, decreases the sensitivity of the tube. To combat this, many tubes are made with PDA (post deflection acceleration) in which a gradually increasing voltage is applied to the electron beam after it has passed the deflection plates, so that the sensitivity is only slightly affected by the extra voltage, but the trace brilliance is greatly increased.

PDA may take the form of a carbon spiral painted inside the tube from a point on the glass bulb close to the deflector plates to another point close to the screen, or it may consist of a metal mesh which acts as a screen between the deflector plates and an accelerating electrode. Both types of PDA tube have advantages: the mesh type allows greater acceleration with very high sensitivity, but spot size is usually larger than in the spiral coating type of PDA tube.

Fig. 8.3 Double gun c.r.t., showing gun construction. (*Photo courtesy of 20th Century Electronics Ltd.*)

High Frequency C.R.T.'s

When c.r.t's have to display frequencies of several hundred MHz or more, some difficulties arise with deflection. Magnetic deflection is practically impossible, as current in deflector coils cannot be changed rapidly enough, and electrostatic deflection becomes difficult due to the capacitance between the deflector plates and the other electrodes which must be charged up and down for each cycle of the waveform being displayed.

This difficulty can be overcome to some extent by mounting the deflection plates directly into the tube, quite separate from the rest of the electron gun, but even this is ineffective at very high frequencies because in the time that an electron takes to pass from the start of a deflector plate to the end the voltage has changed and no deflection can take place. In such tubes several sets of deflector plates are used and connected up as the capacitors of a delay line so that each set of plates has the same voltage on it as the electron passes. This idea is called 'Travelling Wave Deflection'.

Multiple Beams

For many purposes, oscilloscope traces of signals present at different parts of a circuit can best be compared if they are examined simultaneously. This can be done by beam switching an ordinary instrument c.r.t. so that one sweep is carried out with one set of signals applied, and another separated from it, with the other set of signals applied.

Another way of examining more than one trace at a time is to use a multiple beam tube (shown in Fig. 8·3) and, again, there are two ways of doing this. One way is to split the electron beam from a normal gun and pass the split beams into a set of deflector plates, one set for the X direction and two sets (one for each beam) for the Y direction. Usually one plate of the Y set is common to both beams.

A more recent method is to use quite separate guns and deflection assemblies, coupling the X plates so that all the beams are deflected across the screen in the same way but with independent Y plates. As many as eight guns have been mounted into one c.r.t. in this way.

Circuits used with C.R.T.'s

Fig. 8.4 is a block diagram showing the types of circuits used with c.r.t's in various applications. All applications require some sort of amplifier so that the voltages appearing at the Y plates are sufficient to deflect the beam enough to ensure a visible effect at the screen. This amplifier may have to cope with signals ranging from near d.c. (or d.c. itself) to frequencies as high as 1,000MHz (1GHz) in some cases. For service purposes, an all-round oscilloscope Y amplifier may have a bandwidth of d.c.–10MHz; specialised oscilloscopes may have much wider bandwidths. Note that these are not tuned amplifiers but wideband amplifiers.

A simpler wideband amplifier may be found feeding the X plates. Each amplifier will incorporate a phase inverting stage so that signals can be fed to both plates of a pair (if one signal were not inverted there would be no deflection), and also some method of applying d.c. deflection voltages to the plates so that the position of the trace can be shifted to any part of the screen (X and Y shifts).

The X amplifier is fed from a timebase in most oscilloscope applications of the c.r.t. This circuit produces a voltage which deflects the electron beam at a steady speed across the screen and returns it to its starting position very rapidly. During the steady speed (sweep) period, the intensity of the beam may be varied (modulated) by applying signals to the grid of the gun.

In many radar applications, a PPI (Plan-Position Indicator) scan is used. The beam starts at the centre of the tube (representing the radar position) and scans out to the edge. It returns to repeat the process with the scan slightly shifted so that the far end has moved slightly round the edge. In this way, the scan seems to consist of spokes of a circle. The signal is used to modulate the brightness of the tube and the timebase is applied to both sets of coils of a magnetically deflected tube.

Failure of C.R.T.'s

C.R.T. failures are due generally either to old age or to mishandling. Low cathode emission is the most common old age failure; the

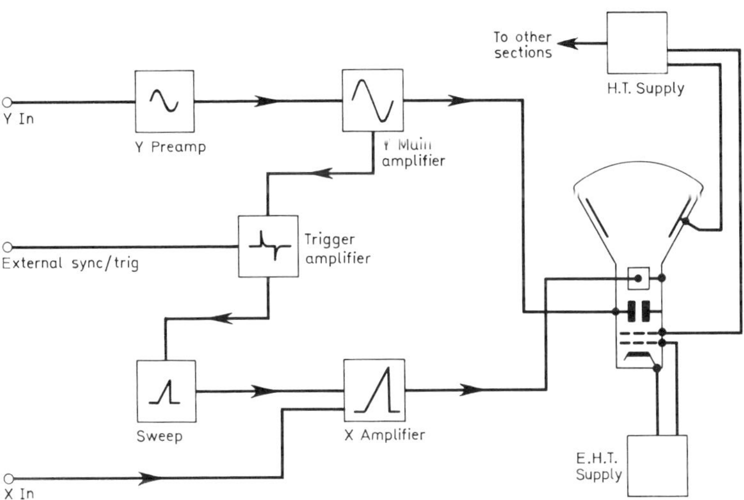

Fig. 8.4 Block diagram of circuits surrounding c.r.t.

brilliance control (G1 bias) has to be advanced further and further to obtain the same brightness of trace under the same conditions, and the flyback (return of the scanning spot) becomes more evident. When this happens, the exact voltage between grid and cathode of the tube should be measured using a high impedance voltmeter (valve voltmeter, MOSFET meter or electrostatic voltmeter) and checked against the makers figures for bias, since the bias network may be faulty.

If the bias is abnormally low, and brilliance is also low, low emission is proven. If the tube cannot readily be replaced, some extra life can be obtained by attaching it to a circuit which will run the heaters only for a minute at 50% extra voltage ($1\frac{1}{2}$ times normal) and then run heaters at 20% extra voltage ($1\frac{1}{5}$ times normal) for an hour with the grid 3–10V more positive than the cathode and passing about 1mA.

The grid current should be checked at intervals, and the power switched off when the grid current ceases to rise. This treatment may give a tube anything from one month to one year extra life as far as emission is concerned, but contains some element of risk, as a tube which is on its last gasp will show less emission afterwards. The risk may be worth taking if the quoted delivery time for a new tube is six months.

Screen burn is another old age sympton, in which the screen becomes darkened in the places where the beam most often strikes

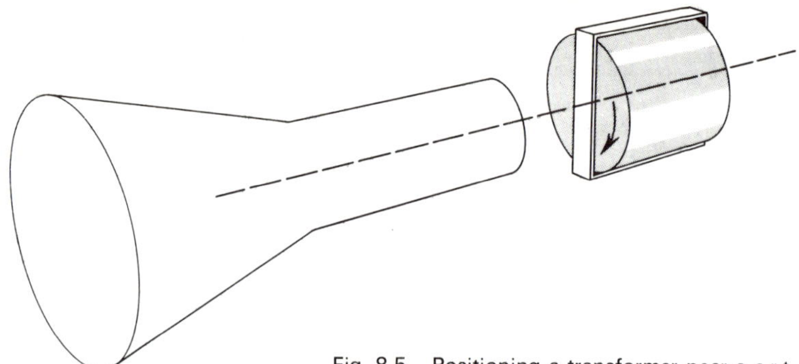

Fig. 8.5 Positioning a transformer near a c.r.t.

it. In oscilloscopes this is usually a horizontal line; in radar tubes a spot in the centre. In addition to this, the efficiency of the whole screen slowly drops, decreasing the maximum brilliance, but this is a long (10 years or so) process.

Damage to the gun structure can cause odd faults such as a misshapen spot (more often due to circuit faults causing the average voltage of the deflector plates to be very different from the voltage of the electrodes near them), disappearance of the spot when deflected (usually displaced deflector places) or complete lack of spot. In every case of suspected damage, all the voltages to the tube should be checked before removing the tube to inspect it, since there is always some risk of damaging the tube while it is being removed.

Mu-metal Screens

Electrostatically deflected tubes used in oscilloscopes and similar types of display can also have their electron beams deflected by magnetic fields such as the field of the earth (which acts as a magnet) or the field magnets, transformers and chokes. Magnetic fields from transformers are usually the worst offenders, since they cause the spot to deflect to and fro so that it appears to be a short line instead of a round spot.

Two precautions can be taken: the tube can be encased in a screen made of mu-metal, which acts to magnetic fields the way blotting paper acts to water, and the tube can be positioned so that the magnetic field of the nearest transformer does not deflect the beam (which means that the axis of the tube should also be the axis of the transformer), or both (Fig. 8.5). Both precautions are usually taken, but great care must be taken with a mu-metal screen. Like any other high permeability material the mu-metal is very easily changed magnetically, and the following rules must be obeyed:

1. No magnet must be allowed near the screen.
2. The screen must not be knocked, tapped, sawn, bent or mechanically jarred in any way.
3. The screen must not be heated violently.

Tuning Indicators

A tuning indicator is a type of miniature c.r.t, in which the beam of electrons, instead of being focused to a spot and deflected in various directions is spread out to various widths by different signal strengths. The construction of a tuning indicator is shown in Fig. 8.6. Most tuning indicators incorporate a triode using the same cathode so that input signals of 10–20V can be used to drive the grid of the triode which then drives the indicator. The circuitry required is also shown.

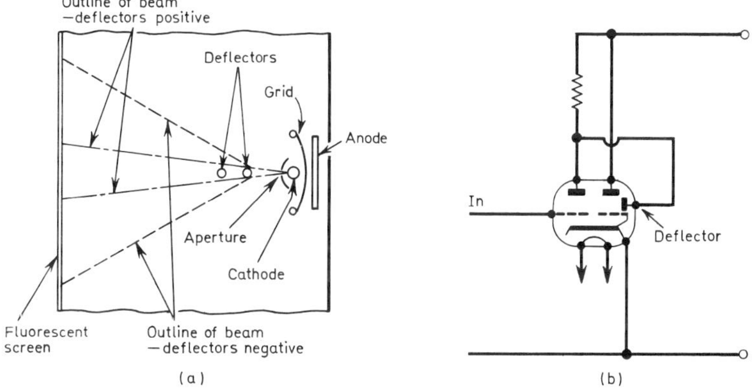

Fig. 8.6 (a) tuning indicator (electron beam indicator), (b) circuit used with indicator.

Gas-filled Valves and Indicators

In the normal thermionic valve, electrons move from the cathode, which is heated to a temperature suitable for efficient emission of electrons, to an anode. This process would take place most efficiently in a perfect vacuum, in which there would be no atoms to obstruct the flow of electrons from cathode to anode. In reality, a perfect vacuum cannot be achieved, but we can pump enough air from the valve to ensure that the remaining atoms do not interfere too seriously with the movement of electrons.

If a triode valve is insufficiently pumped, or if a gas is deliberately introduced, then the effects become noticeable in two ways. Firstly, the grid of the valve begins to lose control of the electron flow, so that negative voltages on the grid no longer cut off the anode current. Secondly, a blue glow can be seen inside the valve. Both of these effects are caused by the action of the electrons on the gas atoms in the valve.

The action of the anode voltage in any valve is to accelerate the electrons, meaning that their speed increases steadily from the instant they leave the cathode until they strike the anode. This, at least, is what would happen in a perfect vacuum, but when the atoms of a gas are present there is a strong chance of electrons striking the atoms.

When a fast-moving electron strikes an atom, the atom breaks up into a positive particle called an ion, and more electrons. If this happens infrequently it makes very little difference to the action of the valve, but if a huge number of ions and new electrons are separated in this way, each one being accelerated by the applied voltage, then the action of the valve changes completely.

For example, the break-up of the gas atoms may be complete, releasing a very large number of electrons and causing the grid action to be ineffective. If the gas is so ionised, then making the grid of the valve negative may cut off electrons from the cathode but cannot affect the electrons created by ionisation. The movement of these electrons in the valve will continue, with more collisions always releasing more electrons, until the accelerating (anode) voltage is removed.

The ions and electrons do not have a long life. Some reach the valve electrodes, the electrons to the anode and the positive ions to the cathode and lose their charges there; others recombine. When the ions of any gas recombine with electrons to form complete atoms, light is emitted; this accounts for the blue glow seen in a 'gassy' valve, and is made use of in gas-filled indicators.

Cold and Hot Cathodes

Both cold-cathode and hot-cathode gas filled tubes are manufactured. As the names suggest the cold-cathode tubes use no heaters, and contain gas (usually a mixture of the gases Neon, Helium and Xenon) at a rather higher pressure than the hot-cathode tubes which use a conventional heater and cathode. The gas pressure used is higher compared to the 'hard' vacuum valve, but low compared to the pressure of the air.

To give an idea of the pressures involved, the pressure of air on a 'normal' day is around 760mm of mercury; the pressure inside a high vacuum valve is around 10^{-5}mm (very much less in some transmitting valves) and pressures ranging between 1mm and 10^{-2}mm are used in gas-filled valves. The word 'filled' is rather unfortunate. The gases used are extracted from the air by freezing, and then distilled to liquid. They are composed of single atoms rather than the groups of two or more found in most other gases and so impart more predictable characteristics to the tubes in which they are used.

At the pressure of gas used in a cold cathode tube, a voltage between anode and cathode causes no effect until a certain minimum voltage, known as the striking voltage, is reached. There are two reasons for this. One is that any electrons present must be accelerated to a high speed if they are to ionise gas atoms by collision, and the speed reached by the electrons depends on the square root of the voltage applied (so that to double the speed of electrons four times the voltage is needed).

The second reason is that electrons have to be obtained from somewhere, and low voltages cannot tear electrons from the cathode surface. This second reason is not quite so important, as some electrons are always being made available by the action of cosmic rays (high speed particles from outer space), but does play some part, as can be seen when a radioactive substance is brought near a gas filled tube. This has the effect of lowering the striking voltage noticeably.

In some cold-cathode tubes, the gas is itself radioactive, emitting electrons which then can be accelerated. The action of light will also help electron emission, so that the striking voltage of a cold cathode tube is greater in the dark than in intense light.

When the tube is conducting, it has an effective incremental resistance which is very low or negative. Negative resistance means that the voltage across the tube becomes less as current through the tube increases. This is a natural effect of ionisation, for more current means more ionisation, more particles to carry current, less resistance and so a lower voltage. Note that the incremental resistance is referred to, meaning the ratio

$$\frac{\text{change of voltage}}{\text{change of current}}, \text{ and not } \frac{\text{operating voltage}}{\text{operating current}}.$$

In any gas filled tube, however, the amount of ionisation is enough to allow very large currents to flow and the circuit must be designed with enough series resistance to make sure that the current is limited in value, or flows for a short time, to avoid damage to the valve.

With the current limited, the voltage between anode and cathode is less than the striking voltage and is termed the 'running voltage', In many cold-cathode valves, this value will lie between 85V and 150V, and it changes very little with change of current.

Hot-cathode valves and tubes have lower gas pressures and have striking voltages which are very high, several kV. The electrons are provided by a hot cathode and controlled by a grid, so that the valve can be switched suddenly into conduction. Once conducting, the valve passes current until the voltage between anode and cathode becomes too low. If the voltage remains low until all the ions have recombined, the valve will not conduct when the voltage is raised again until the grid is pulsed to allow more electrons to restart ionisation. This is the principle of the thyratron.

Neon Indicators

Small neon indicators used in electronics may be used purely as panel indicators, or may play a part in a circuit. As the voltage across the neon gas discharge is roughly 85V over a fair range of currents, these indicators can be used as voltage stabilisers when the need is not for very accurate stabilisation.

They can also be used as indicators of voltage level, indicating by the current flow through the neon when the voltage across the neon has reached the striking voltage of the neon. A use based on the same principle is a timer, where a capacitor is charged until the voltage is enough to strike a neon; the time taken to charge the capacitors is fairly constant if the charging is through a large resistance value for a fairly high voltage, several times the striking voltage.

If the capacitor in this circuit (Fig. 8.7) is not discharged by other means (such as a relay contact), then a sawtooth waveform is generated by the discharge of the capacitor through the neon, eventually resulting in the voltage going below the level where discharge can take place, followed by charging again to the striking level.

Voltage Stabilisers

The stability of the running voltage of a gas filled tube can be greatly increased by suitable design, notably in the shape of anode and cathode and the materials used. It is also possible, by using different gases and mixtures of gases, to make stabilisers of different running voltages. Such voltage stabilisers may be made physically small if they are to handle low currents, and the stabilised voltages

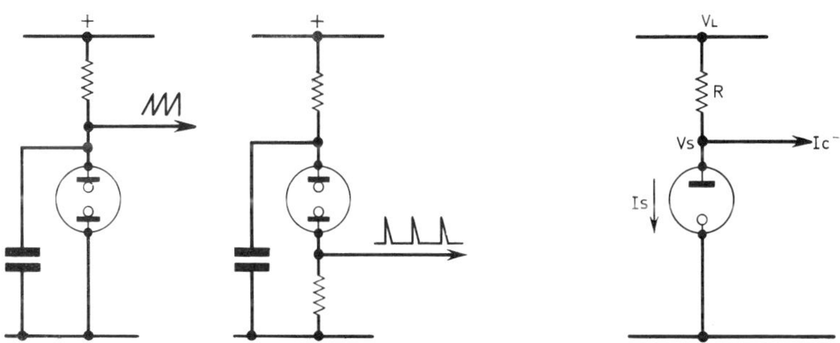

Fig. 8.7 (left) Circuits for neon indicators or glow diodes.
Fig. 8.8 (right) Basic circuit for voltage stabiliser.

are rather higher than can conveniently or cheaply be obtained directly from Zener diodes (see later). Both wire-ended and plug-in versions are obtainable.

In use, gas stabilisers have a fairly large voltage difference between the running voltage and the maximum possible striking voltage. For example, a stabiliser with a running voltage of 86V may have a maximum striking voltage of 125V, so that the operation of an 86V stabiliser requires the provision of a supply at a voltage greater than 125V. Unless the gas filling is radioactive, as noted earlier, the striking voltage in conditions of darkness may be greater still. This discontinuity is not found with Zener diodes.

In use, stabilisers are connected in parallel with the circuit whose voltage is to be stabilised (Fig. 8.8). The supply is led in through a series resistor whose value can be calculated knowing the maximum current demand of the circuit to be stabilised. If the current taken by this circuit is I_C, and the minimum current which will keep the stabiliser tube ionised is I_S, then the resistor must pass a total current of $(I_C–I_S)$ at a voltage which is the difference between the supply voltage V_L and the stabilised voltage V_S. By Ohm's Law, the resistor value is $(V_L - V_S)/(I_C - I_S)$ ohms.

In practice, a rather larger current would be passed to avoid the stabiliser being de-ionised after a surge of current, so that the next preferred value of resistor below this calculated value would be used. The wattage dissipated in the resistor is $(I_C–I_S) \times (V_L–V_S)$ and the wattage dissipated in the stabiliser varies from $V_S \,.\, I_S$ when the circuit takes its maximum current to $V_S \times (I_C–I_S)$ when the circuit takes no current. Care has to be taken in the design of stabilised circuits that these conditions are observed.

If the degree of stabilisation required is greater, usually for use in a more elaborate voltage stabiliser circuit, then voltage reference tubes can be used. These are stabilisers whose design is such that the change of voltage for a given change of current is very much less than for the conventional stabiliser. In addition, the voltage change for change of temperature may be considerably less than for a stabiliser.

Alpha-numeric Indicators

The two features of gas-filled cold-cathode tubes which we have seen applied so far (the flow in the gas due to recombination of ions and the difference between striking and running voltage) are both made use of in alpha-numeric indicators. The glow in any gas-filled tube run on d.c. is concentrated round the cathode, as this is where large numbers of positive ions are heading and can meet the electrons so that recombination takes place.

If the cathode of such a tube is made of wire bent in the shape of a figure or a letter of the alphabet then the glow surrounding it when current passes will be of the shape of the wire. It is possible to make tubes containing a complete set of cathodes in the shape of the numerals from 0–9, one behind the other and each separately connected to a base pin. When one cathode is made sufficiently negative, then the gas discharge current starts and the cathode glows.

We can make use of the difference between running voltage and striking voltage to reduce the signal voltage which we need. If the cathodes are connected to earth through a resistance and the anode of the tube is taken to a voltage rather lower than the striking voltage, then a negative pulse applied to a cathode, sufficient to make the voltage between that cathode and the anode greater than the striking voltage, will cause a discharge to that cathode. The discharge will continue after the end of the pulse, so that the number remains displayed.

Switching off could be effected by pulsing the cathode positive (below running voltage) but is most easily done by pulsing the anode negative. Tubes of this type can be obtained showing a wide variety of numbers, letters and symbols; they can be operated with a roughly smoothed half-rectified a.c. supply if necessary.

Trigger Tubes

Trigger tubes are gas-filled cold cathode triodes in which to the normal anode and cathode has been added a third electrode, the trigger electrode, very close to the cathode and consisting of a pointed wire directed to the cathode. Because of the shape of this trigger electrode and its nearness to the cathode, a gas discharge

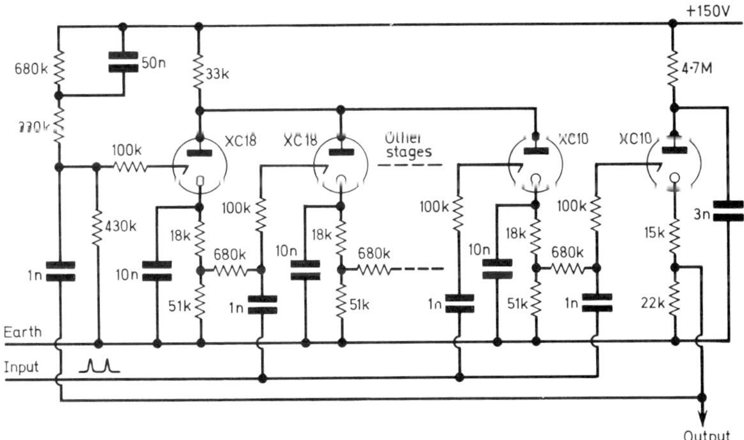

Fig. 8.9 Ring counter using trigger tubes. The number of tubes in the ring determines the number of input pulses needed to produce one output pulse.

between trigger and cathode can take place at lower voltage than the normal discharge between anode and cathode. Once the discharge has started between trigger and cathode, however, the ions created will cause current to flow also between anode and cathode, provided that the voltage is high enough.

A trigger tube, therefore, passes no current until a pulse at the trigger electrode starts it off. The voltage and duration of the trigger pulse must be enough (typically 100V and 75μS) to ionise the gas and start the main discharge, but once the main discharge has started it will continue no matter what voltages are applied at the trigger electrode. The main discharge can be stopped only by interrupting the anode or cathode supply or by lowering the anode-to-cathode voltage below the running voltage.

Cold-cathode trigger tubes are extremely useful, therefore, in counting and gating circuits, particularly if fairly high voltages of pulse outputs are needed. The most popular counting arrangement is the ring counter (Fig. 8.9), in which a number of trigger tubes are connected as shown, with the d.c. supply to each trigger electrode taken from the cathode of the previous tube except for the first tube in the ring.

In operation, with one tube switched on, only the trigger electrode of the next tube in line will be at a high enough voltage to trigger when the pulse input is taken to all of the trigger electrodes in the ring. When the pulse arrives and causes the next tube to conduct the voltage drop across the common anode supply resistor switches off the previous tube. The cycle can then take place again.

Some care has to be taken in the design of such circuits; the capacitors between the cathodes and earth prevent the cathode voltage changing too rapidly when current starts to flow, otherwise one pulse would trigger off all the tubes. The size and duration of the trigger pulse must be carefully controlled in any cold-cathode circuit, otherwise trouble can be expected. The counting rate is limited by the time taken for de-ionisation of the gas, and by the need to limit the rate of rise of cathode voltage and is usually low (1kHz).

Glow Diodes

Glow diodes are miniature (about 2cm long) cold-cathode discharge tubes which are particularly well suited for use in modern transistor circuitry. Though they are similar in appearance to neon indicator lamps, their construction is aimed at stable operation rather than light output and some types can be used for voltage regulation, with temperature coefficients of about 1·5mV per °C; rather less than that obtainable with Zener diodes. Typically, these reference glow diodes can have stabilised voltages of 82V to 100V at currents of a few mA, all with negative temperature coefficients.

A wider range of applications for general purpose glow diodes is found in timing and triggering circuits. Tubes are available with striking voltages ranging from about 70V to about 400V, with running voltages ranging between 50V and 180V. Typical uses include timers, photochoppers, proportional controls, and transient suppression. Timing circuits use the difference between striking and running voltages to establish a sawtooth voltage waveform. The sudden current through the tube can be used to operate a relay, or a pulse can be taken from a series wired resistor; alternatively a photocell can be activated from the light output.

Chopping circuits convert d.c. voltage levels into squarewaves by switching an output from the d.c. level to earth level, but it is always difficult to isolate the switching waveform from the voltage being chopped, which is usually small. Using a photo-resistor as a series or shunt element with glow diodes performing the switching gives very considerable isolation.

Proportional controls enable devices such as a.c. electric motors and fluorescent lamps to be adjusted to different output levels. In this application, glow diodes can be used as triggers for the thyristors which form the power control devices. Used in transient suppression circuits, glow diodes can be wired across relays or other inductive loads to absorb the large reverse e.m.f. which is generated when such devices are switched off.

For a comprehensive survey of neon discharge devices and suggested circuits, the reader should consult *Applications of Neon Lamps and Gas Discharge Tubes* by Edward Bauman (Carlton Press, N.Y.)

Hot-Cathode Gas Valves

Gas-filled rectifiers use a much lower gas pressure than is found on cold-cathode tubes, and rely on a hot cathode for their electrons. The ionisation which takes place when the anode is positive is fairly complete, so that large current can flow. In addition, the forward resistance is low. Because the electron supply comes from the cathode, the current ceases when the anode is negative, and because the gas pressure is low the negative voltage on the anode can reach a high value without any risk of creating ions which would cause conduction.

The gas-filled rectifiers currently available use either mercury vapour or Xenon as the gas filling. Mercury vapour is used in the smallest valves available and is available in the larger sizes, while Xenon tends to be used in the larger valves only. Where mercury vapour rectifiers are in use, the heaters must be run at full operating temperature before the anode voltage is applied, and a suitable time delay circuit must be used to achieve this.

The reason is that the mercury condenses into pools of liquid at room temperature, and must be vaporised before the valve can operate as a gas-filled rectifier. Using the valve too quickly means that the cathode has to supply the forward current with no help from mercury ions, and this is usually destructive to the cathode.

Thyratrons

These are the big brothers of trigger tubes. Small thyratrons are made in glass envelopes and resemble power output valves; the larger sizes have ceramic cases and look like top hats. Their circuit action resembles trigger tubes inasmuch as a pulse applied to a grid electrode can start the gas discharge between anode and cathode, and this can be stopped only by interrupting the current or lowering the anode voltage to below the sustaining voltage.

Unlike trigger tubes, thyratrons have hot cathodes and low gas pressures. The voltage which can be applied between anode and cathode is very much larger than can be handled by any other triggering device, ranging from 1kV for the smallest thyratron to 120kV for the largest. The currents, which can be passed are also large, up to 6,000A in the large ceramic-cased thyratron. Pulse power outputs of up to 200MW can be obtained.

Thyratrons are used in control circuits (notably spot welders), radar transmitters (to switch magnetrons), in large inverters (converting d.c. to a.c. by switching + and − supplies alternately to the primary of a transformer) and to operate other high-energy short-time devices such as flash tubes. In many cases, the lower voltage and current applications are now carried out by the solid-state equivalents, thyristors.

CHAPTER NINE

SEMICONDUCTORS I

MATERIALS WHICH CONDUCT do so because they have large numbers of electrons available and free to move. Heating such materials drives off those available electrons and decreases the conductivity (in other words, increases the resistance). This is the normal behaviour of a metal conductor. Materials which are insulators have no available electrons to conduct; all their electrons are bound tightly to the atoms of the material and can be set free only at very high temperatures, usually only when the material is molten.

Between these categories, there are materials which have a few available electrons for conduction, whose numbers can be greatly increased by heating or by adding impurities which are rich in electrons, or, oddly enough, deficient in them. These are the substances which we call semiconductors and which have completely changed the whole of electronics since the announcement of the discovery of the transistor in 1948.

The Hole Theory

In semiconductors, two methods of carrying current exist—one in which electrons move steadily, and another which involves electrons shuffling position to fill up a gap where an electron has been removed (the gap is usually called a 'hole'). If the electron stream is thought of as a normal line of cars in a traffic stream, 'hole' conduction is similar to the case when a gap in the traffic is filled up by a car accelerating, then by the next one following it, and so on (Fig. 9.1).

The gap appears to move in the opposite direction from the cars and at a slower speed (because each car takes longer to close up to the one in front than it would if the one in front were stationary). So with semiconductors, the conduction due to 'holes' seems to be caused by slow moving particles.

Summing up: semiconductors are materials which conduct

poorly at room temperature unless impurities are added (a process known as doping); they conduct by electron flow and by electron 'shuffling' (hole current); their conductivity approximately doubles for each rise of 20°C in temperature.

Semiconductor Diodes

When a crystal of semiconductor material is doped with different impurities from each end so that one end has a surplus of electrons and the other has a surplus of holes, and the two zones meet inside the crystal, a semiconductor diode has been formed. The semiconductor with a surplus of electrons is called n-type, that with the surplus of holes is called p-type, and the zone where they meet is the junction. Junctions must be part of the crystal. They can be grown by allowing molten p-type material to crystallise on to a n-type crystal, diffused by letting one type of material spread into another by heating, or formed by epitaxy, which involves letting the vapour of one type condense on the crystal of another.

Fig. 9.1 'Movement' of a gap, caused by attempts to fill the gap. This simple explanation of the hole is only intended to give an impression of how a hole can simulate a particle. The true picture is much more complicated, as the hole appears to have mass, charge and velocity. In all respects, within a crystal the hole is as real as the electron; it has no existence, however, outside the crystal.

When the p-type portion of a junction has a positive voltage applied to it and the n-type portion has a negative voltage applied, electrons are attracted to the positive side and holes to the negative side: current flows (Fig. 9.2). When the voltages are reversed, electrons are attracted away from the p-region and holes from the n-region, leaving the junction stripped of charges and unable to conduct except by the few electrons and holes liberated by heat. The semiconductor is now behaving as a diode, and passing current readily in one direction but not in the other. Such an arrangement is called a semiconductor diode, and can be formed of any material which is semiconducting.

The choice of a semiconductor diode for a particular job must be

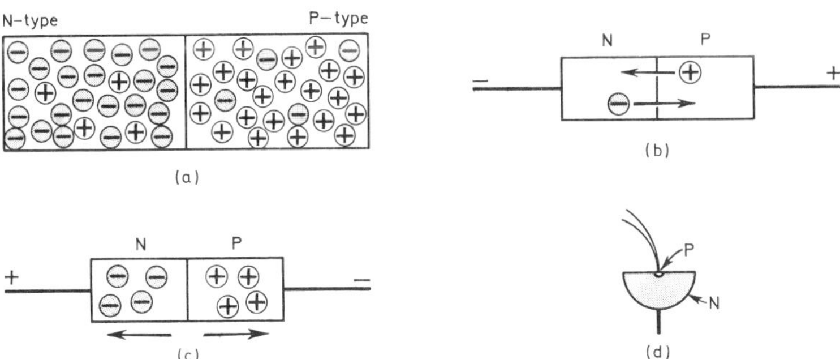

Fig. 9.2 Junction. (a) note that the carriers marked are the *free* charges. Both p and n-type semiconductors do not contain extra charges; they are electrically neutral. What the p-type contains is a greater number of *free holes* compared to free electrons; the n-type contains more free electrons than holes. In each case the total number of electrons (free and bound) equals the total number of holes, free and bound; (b) Forward-biased diode: carriers move across the junction; (c) Reverse-biased diode, junction is stripped of carriers; (d) Point-contact diode —a small junction is formed round the point of contact.

made with care, as the variation between types is as great, though not so obvious to the eye, as the variation between different types of valve diode. To start with, there is a choice of several types of semiconductor material (whereas valves have no choice of different types of vacuum!), although in practice, germanium and silicon are the main contenders. The type of structure is also important; point contact diodes consist of a block of one type (p or n) of semiconductor with a contact spike forming an impurity contact at which rectification takes place, whereas, in a junction diode, the meeting place of types is within the crystal.

Characteristics of Semiconductor Diodes

The most important figures usually quoted for semiconductor diodes are:

(1) The peak reverse voltage—the maximum voltage which can be applied to the diode in the non-conducting direction without causing breakdown of the semiconductor material. This peak reverse

voltage (+ or −) depends on the type of construction and material used, but is always much less than the maximum attainable with thermionic diodes (several hundred thousand volts). The PRVs found vary from about 10V up to several thousand volts for silicon rectifiers; thermionic diodes have PRVs ranging from about 100V to the region of 100kV for rectifiers used in power supplied for X-ray sets and high power transmitters.

(2) Maximum forward current. This is the maximum average current which may be allowed to flow through the diode when it is conducting. Very often a peak forward current may be quoted as well, and the distinction between peak and average is important. If a current of 100A flows for 1mS every second (pulse of 1mS width with a PRF of 1S.), then the peak current is 100A and the average current is 100 × 1mS/1000mS = 0·1A.

Each is important, as the average current determines the temperature rise in the whole diode, and the temperature rise causes the diode to have a lower PRV. The peak current causes local heat at the junction or point of the diode and may cause evaporation of material with loss of diode action.

(3) Voltage drop and minimum forward voltage are related figures. The minimum forward voltage is the forward (conducting) voltage across the diode for some specified current (frequently 1mA). The voltage drop is the voltage across the conducting diode at another current or current range. These figures indicate what minimum voltage can be rectified and what power is dissipated in the diode at a given current. The power dissipated while current is flowing is: (voltage drop × current). The voltage drop and minimum forward voltage may affect the use to which the diode is put. Too high a voltage drop may make a diode unusable in a gate circuit; too high a minimum forward voltage causes nonlinearity in a detector stage.

(4) Reverse leakage current. In any diode, a small amount of current flows in the reverse direction when the diode is reverse-biased. This reverse current is small, but increases with increasing temperature. The greater the forward current rating, the greater the reverse current, because reverse current is determined by the same factors of structure as determine forward current.

(5) Stored charge. When current flows in a semiconductor diode electrons move into regions mainly occupied by holes, and holes move into regions mainly occupied by electrons. When the forward

voltage is removed, the electrons and holes do not instantly disappear from the regions where they are in the minority, and if a reverse voltage is applied very quickly, there may be a considerable current in the reverse direction before the excess holes and electrons are used up.

This affects the rate at which the diode can be switched from a forward voltage to a reverse voltage, and also affects the operation of the diode at high frequencies. The amount of 'minority carriers', as these electrons and holes are called, which are still 'in transit' when voltage is reversed, can be measured, and the figure produced is termed 'stored charge', usually quoted in unit of charge of picocoulombs (pC, equal to the amount of charge transferred when one microamp flows for one micro-second).

The first letter indicates the type of semiconductor material used.
A . . . Germanium junction
B . . . Silicon junction
C . . . Gallium Arsenide or similar compounds forming junction.
D . . . Indium Antimonide or similar compounds forming junction.
R . . . Photoconductor, or other device with no junction.

The second letter indicates the construction or use of the device.
A . . . detector diode, high speed diode, mixer diode.
B . . . variable capacitance diode.
C . . . transistor for a.f., not power amplification.
D . . . Power transistor for a.f.
E . . . tunnel diode.
F . . . transistor for r.f. amplification (not power).
G . . . multiple transistor with dissimilar devices on same chip.
L . . . power r.f. transistor.
P . . . photosensitive device or other detector of radiated energy.
Q . . . radiation emitting device such as light-emitting diode.
R . . . thyristor or other switching device with specified breakdown characteristic. Not high power.
S . . . switching transistor, not power types.
T . . . thyristor or other controller, high power.
U . . . power transistor for switching.
X . . . multiplier diode such as varactor or step recovery diode.
Y . . . rectifier diode, booster diode, efficiency diode.
Z . . . voltage reference or voltage regulator (Zener) diode.

The figures or letters following indicate the design; A three figure serial number is used for 'consumer types', used in domestic radio, TV, tape recorders, audio etc.
A serial consisting of a letter (Z, Y, X, W, etc.) followed by two figures means a device for professional use (transmission etc.)

Table 9.1—PRO-ELECTRON CODE FOR SEMICONDUCTORS

Types of Semiconductor Diodes

Basic Form: *Point contact diodes* have very small active areas, and therefore small amounts of stored charge.They cannot pass high currents (usually limited to about 100mA), nor can they withstand high reverse voltages (usually limited to about 100V). Point contact diodes are cheap and easy to make, and can be used at high frequencies. *Junction diodes* have larger active areas and pass higher forward currents. They are extensively used for power rectification, and the junction may be made to withstand high reverse voltages (up to several kV).

Materials: *Germanium* has higher conductivity than *silicon*, increases its conductivity more rapidly with increasing temperature, and breaks down completely at lower temperatures than silicon. Diodes made from germanium have lower voltage drops and lower reverse currents than diodes of similar construction made from silicon.

Construction: Various methods of construction may be found, mainly to reduce stored charge. Junctions formed by the evaporation of one layer of semiconductor over another are called *epitaxial*, and this technique (used also in transistors) gives accurate control over the dimensions of a junction so that stored charge may be made very small.

Failure of Semiconductor Diodes

Semiconductor diodes may be found in any circuit application mentioned as applicable to thermionic diodes, except for the rectification of very high voltage. Unlike thermionic diodes, which eventually lose emission, there seems to be no natural 'wearing-out' process for semiconductor diodes, and they continue to function unless damaged in any of the following ways:

(1) Overheating: If the diode is operated at more than its maximum rated temperature (usually 70°C for germanium types; 100°C for silicon) at full rated PRV and forward current, then the p–n contact area is damaged beyond repair.

(2) Overloading: If PRV or forward currents ratings are exceeded, then again, the p–n contact area is damaged by internal overheating. This may happen even if the excess current passes for less time than is needed for a fuse to blow; because of this, other forms of protection are often needed in addition to fuses.

A diode may fail by short circuiting or open circuiting. Usually, excess forward current causes open circuit failure and excessive

PRV causes short circuit failure (which may go open circuit if a large reverse current flows).

Under correct operating conditions, diode failure is practically unknown, and the same may be said of transistors, with the possible exception of very high frequency, high power types.

Transistors

If a thin wafer of p-type germanium or silicon has n-type regions formed on opposite sides, one might expect this combination to behave like two diodes connected back to back, and indeed it does. This construction has some remarkable additional features, however, for it is in fact a transistor. Because the p-type region is so thin, the two diode junctions are not isolated, and electrons moving through one junction can also pass through the other rather than changing direction, in the same way as a car on a greasy road may skid straight on instead of turning a corner.

Referring to the diagram of Fig. 9.3. suppose that leads can be taken to each portion of the transistor. If one n-type region is made +, and the other −, we are ready to demonstrate transistor action. Suppose the p-region is now made slightly positive. By normal diode action, electrons should now flow from the negative n-region to the p-region, since this is a forward biased diode.

This certainly happens, but very few of these electrons 'turn the corner' and end up flowing along the leads from the p-region; instead, most of them simply pass straight through the thin p-region, over the other p–n junction (easy enough, because they are

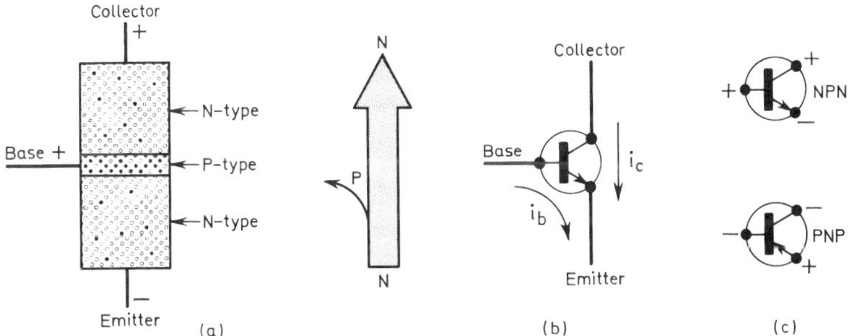

Fig. 9.3 (a) Action of junctions in a transistor. Direction of electron flow is shown; (b) Symbol for transistor, showing directions of currents shown in the conventional way (positive to negative); (c) The direction of the arrow in the transistor symbol is used to show the type of polarity.

moving rapidly) and into the + connected n-region and so to the external circuit.

Typically, the current flowing straight through in this way is 100 times or more the current which 'turns the corner' and flows into the circuit connected to the p-region. If the p-current is 1mA, the n-to-n current is 100mA, but what is more important is that if we double the p-current (by making the p-region more positive), the n-to-n current is also doubled. This accounts for the transistor's action as a current amplifier.

To avoid having to refer to p-region and n-region, we have definite names for the parts of a transistor. The thin portion sandwiched between the junctions is the *Base*, the region emitting current to the base with a forward bias is the *Emitter*, and the region which collects the current which goes through the base is called the *Collector*. Transistors may be classed as p-n-p or n-p-n according to the base material and the junctions formed on it.

The majority of germanium transistors are p-n-p; the majority of silicon transistors n-p-n. In n-p-n transistors, conduction in the base region is by electrons from the emitter as we have discussed; in p-n-p transistors, conduction in the base region is mainly by 'holes' and the connection to all the electrodes are of opposite polarity to these used for an n-p-n transistor.

Note carefully that it is the *current* between base and emitter which causes a larger *current* to flow between collector and emitter. In the valve, it was the *voltage* between the grid and the cathode which controlled the *current* between anode and cathode.

Parameters of Transistors

The 'parameters' of any amplifying device are a set of figures used to describe the performance of the device. We have previously seen the valve parameters g_m and R_a; they are reasonably simple and easy to use because the comparatively long history of valves has enabled us to dispense with parameters which are of no further use. Transistors, being much more recent have been cursed with a huge conglomeration of parameters because each different manufacturer tended to use sets of parameters which showed his products in the most favourable light. The parameters more frequently used now are:

hfe . . . . the current amplification at 1kHz with input signal fed between base and emitter and the output signal taken from the collector.

fhfe , . . . the frequency at which the current amplification hfe has fallen to 0·707 of its value at 1kHz (−3dB).

fr . . . . hfe × frequency (at frequencies of fhfe and more).

f_1 . . . . the frequency at which hfe (using a resistive load) has fallen to 1.

*I*ceo . . . . Leakage current with collector reverse biased and base o/c.

*I*cbo . . . . Leakage current with collector reverse biased and emitter o/c.

*I*ces . . . . Leakage current with collector and emitter reverse biased.

*I*cbs . . . . Leakage current with collector reverse biased and base s/c.

*V*cbo . . . . Maximum collector-to-base voltage with emitter o/c (this is the p-i-v emitter o/c).

*V*ceo . . . . Maximum collector-to-emitter voltage with base o/c (this is the p-i-v base o/c).

$I\mathrm{cm}_{max}$ . . . . Maximum peak current in collector.

$P_{tot\ max}$ . . . . Maximum power dissipated at collector.

$T_{j\ max}$ . . . . Maximum permitted temperature at collector–base junction (This is not measurable; the temperature of the case will be less than this).

Most of these parameters are of use only to the designer, as are also the remainder of the parameters which are concerned with the performance of the transistor in circuits which work it to its limit.

Equivalent Circuits

Because of the shorter history of transistors, a very large number of equivalent circuits have been introduced, used and discarded in the years since 1948. Most of the early equivalent circuits were made necessary because of the imperfections of the transistors which were then available. Modern transistors have developed to the stage where the current flowing in the base is so small (almost comparable with the grid leakage current in early valves) that it can be almost ignored, and the transistor treated as a voltage amplifier. An article by A. J. Blundell (*Wireless World*, June and July, 1971) has helped to crystallise this new attitude, and serious students of transistor circuit theory are strongly recommended to this article and others quoted in its text.

For most work which does not involve more than understanding

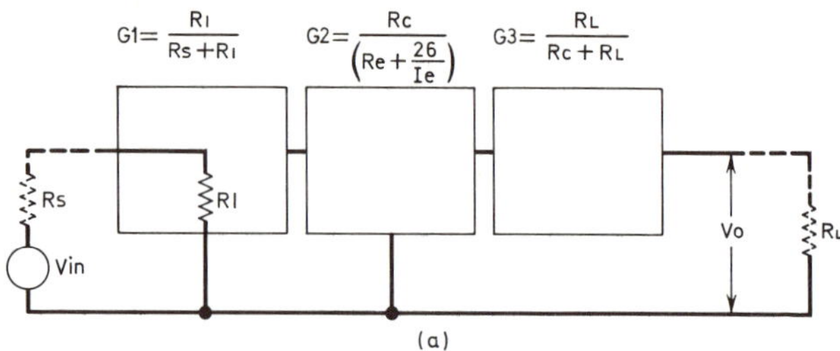

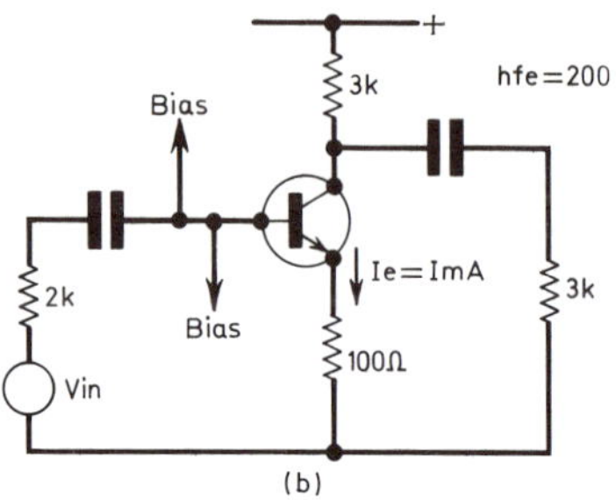

Fig. 9.4 (a) The three-box equivalent. This is usable for common emitter connections at small currents; some 90% of small-signal applications; (b) The three-box equivalent applied to the gain of a voltage amplifier.

(a) The three-box equivalent. This is usable for common-emitter connections at small currents; some 90% of small-signal applications.

$$\frac{V_o}{V_{in}} = G1 \times G2 \times G3$$

$R1 = (26/Ie + Re) \times hfe$, where $Ie =$ emitter current in mA, $Re =$ emitter resistor (undecoupled), and $hfe =$ current gain (common emitter). $Rc =$ collector resistor (from collector to decoupled power supply), $R_L =$ load resistor (a.c. load), and $Re =$ source resistance.

(b) The three-box equivalent applied to the gain of a voltage amplifier.

$$R1 = \left(\frac{26}{1} + 100\right) \times 200 = 25{,}200$$

$$G1 = \frac{25{,}200}{25{,}200 + 2{,}000}, \text{ about } 0{\cdot}93 \quad G2 = \frac{3{,}000}{126}, \text{ about } 24 \quad G3 = \frac{3{,}000}{3{,}000 + 3{,}000} = 0{\cdot}5$$

$$\frac{V_{out}}{V_{in}} = 0{\cdot}93 \times 24 \times 0{\cdot}5 = 11 \text{ (approx.)}$$

If the 100Ω resistor is now decoupled, so that it plays no part in the gain, then:

$R1 = 26 \times 200 = 5{,}200$

$$G1 = \frac{5{,}200}{5{,}200 + 2{,}000} \qquad G2 = 26 \times 200 \qquad G3 = \frac{3{,}000}{3{,}000 + 3{,}000}$$

$$\frac{V_{out}}{V_{in}} = 41 \text{ (approx.)}$$

circuits and doing routine design calculations, a very much simplified transistor equivalent can be used (Fig. 9.4a). This equivalent splits a transistor into three boxes, and the overall gain is the product of the gains in each box (two of the 'gains' being, in fact, losses).

The first of the three boxes is the input box, whose gain is given by $R_1/(R_s + R_1)$ where R_1 is the effective input impedance of the transistor. The second box is the gain box, whose gain is $R_c/(R_e + r_e)$ where r_e is the 'internal emitter resistance'. The third box is the output box, whose gain is $R_L/(R_c + R_l)$ where R_c is the d.c. load resistor and R_c is the load presented by the next stage. The quantities r and r_e depend on the conditions of use of the transistor. Shown in Fig. 9.4b is this equivalent circuit applied to the gain of a voltage amplifier stage.

One notable point which greatly simplifies the assessment of any transistor circuit is that many designers choose a collector voltage of half supply voltage. In such a case the open circuit voltage gain (no load) is about twenty times the supply voltage, and this represents the maximum gain of a transistor stage designed for linear amplification.

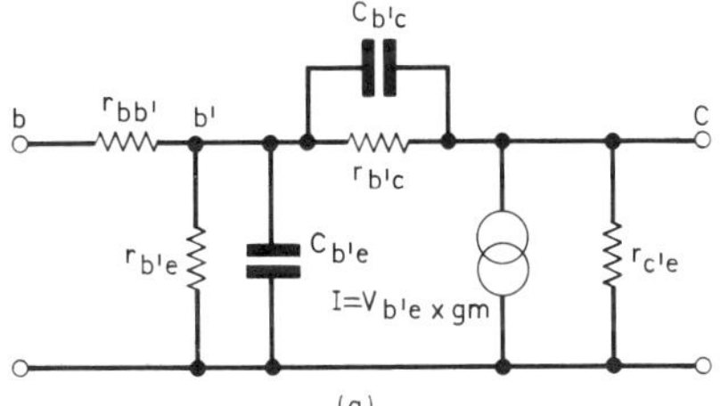

Hybrid-pi parameters measured at Vce = −6·0V, IC = −1mA, frequency 1·6kHz (typical values)

rb′	180Ω
gm	39mA/V
rb′e	700Ω
rc′e	125kΩ
rb′c	3·5MΩ
Cb′c	36pF
Cb′e	10nF

Fig. 9.5 (a) Hybrid-pi equivalent circuit. The letters b. c and e denote the base, collector and emitter terminals of the transistor; (b) examples of hybrid-pi parameters for Mullard BCZ11 (silicon p-n-p general purpose transistor).

The hybrid-pi network is more widely quoted by manufacturers and is shown in Fig. 9.5. It consists of a potential divider into the base circuit, with capacitance to account for the input line constant. This is a feedback network which reduces the gain obtainable and causes high phase shifts at high frequencies (to account for fr) and a current generator with a resistor in parallel. The output of the current generator is $g_m V_{be}$ where g_m is the mutual conductance in mA/V as for a valve. For transistors, the g_m is 39mA/V at 1mA collector current.

Thermal Characteristics

All circuit elements which have resistance must dissipate power; this applies to resistors, valves, transistors or anything else. In a valve, the 'resistance' is in the vacuum between grid and anode, and the power appears at the anode, which heats up. This heat is enough to raise the temperature of the anode to more than the temperature at which water boils (100°C). In power valves, the anode temperature may be 300°C or more, and transmitting valves frequently are operated with the anodes glowing red at 650–700°C. This heat is dissipated by radiation through the glass of the valve, or by cooling the anode by means of fins placed in a current of air, or by a water jacket.

In a transistor, however, the power is converted to heat in the semiconductor material and is *conducted* to the collector. Semiconductor materials are alloys of germanium or silicon with other materials, and their behaviour depends critically on their composition. Excessive heat can cause the other materials to move, thus destroying the junctions on which transistor behaviour depends.

The maximum tolerable temperature at the collector-to-base junction of a germanium transistor is about 90°C, and the figure for a silicon transistor is about 150°–170°C. The amount of power which can be dissipated in a transistor depends therefore on how fast the heat can be carried away from the collector-to-base junction.

When the collector of the transistor is connected to the outer casing, as is the case with power transistors, it is possible to dissipate the heat at the collector junction by conducting it to a metal surface, possibly the chassis of the equipment if this is at collector potential, or directly to a heat sink, which is a sheet of metal, usually provided with cooling fins and painted black.

Where the transistor is cooled in this way, it is important that it should be well bolted down, with a film of silicone grease (which helps the process by filling any gaps with heat-conducting grease) between transistor and mount. Slack mountings, missing grease, missing black paint or broken fins can cause a transistor to fail through overheating.

Where the transistor must be electrically isolated from the heat sink (and especially where more than one transistor is mounted on a heat sink), a thin mica washer, coated with silicon grease on each side, is inserted between the transistor and the heat sink, and the mounting bolts have mica washers and plastics sleeves to avoid shorting the transistor to the heat sink. Very great care should be taken when transistors mounted in this way are replaced. Smaller transistors, using cases isolated from the collector, can be strapped to chassis or have a fin attached.

The calculation of the junction temperature, when the power dissipated in the transistor is known, is fairly straightforward provided that all of the figures required are known. The important quantities are the 'thermal resistances', symbol θ, from the junction to the air into which the heat is finally dissipated. Thermal resistances are quoted in units of °C per watt or °C per mW. If a transistor is quoted as having θ_j-case of 40°C/W, the implication is that with 1W dissipated, the junction will be 40°C warmer than the case, at 2W dissipation 80°C warmer, etc. The thermal resistance from the junction to the case is only part of the heat 'circuit', however, and the heat must then travel from the case to the heatsink and from the heatsink to the air.

Thermal resistances, like electrical resistances, add simply in series, and if the overall thermal resistance from junction to air (which is the important figure) is to be known, then the resistance from case to heatsink and from heatsink to air must be known. These resistances are not always so easily obtainable, though they can be measured. When small transistors, for which a standard clip or heatsink is available, are used, the manufacturer may quote the thermal resistance direct from the junction to air.

For larger transistors, it is usual to assume for a mica insulator a resistance of 0·5°C/W and for a directly bolted on transistor 0·2°C/W between case and heatsink. The thermal resistance between heatsink and air is the most variable feature, difficult and inaccurate to calculate and varying with circumstances. One frequently quoted figure is of 2·2°C/W for a 7 in × 7 in. × $\frac{1}{16}$ in. blackened aluminium heatsink.

The power dissipation of a transistor is limited mainly by the rate at which heat can be removed, so a formula for calculating this is useful. This is:

$$P_{tot} = \frac{T_{j\,.\,max} - T_{amb}}{\theta_{j-c} + \theta_{c-h} + \theta_{h-a}}$$

where $T_{j\,.\,max}$ is the maximum allowed junction temperature,
T_{amb} is the maximum likely operating air temperature,
θ_{j-c} is the thermal resistance, junction to case,
θ_{c-h} is the thermal resistance, case to heatsink,
θ_{h-a} is the thermal resistance, heatsink to air.

For example, consider a large germanium transistor whose maximum junction temperature is 90°C, to be used in an air temperature not exceeding 35°C. If the θ_{j-c} is 2·5°C/W, θ_{c-h} is 0·5°C/W (mica) and the heatsink resistance to air is 2·5°C/W, then the total maximum power which can handled is: (90–35)/(2·5 + 0·5 + 2·5), which equals 10W.

Sometimes, the equation has to be worked in the reverse direction and the temperatures at various parts have to be calculated from the power dissipated. For example, if the transistor in the previous example is to dissipate 7W, then we can find what air temperature will cause a junction temperature of 90°C. Using the same equation: $7 = (90 - T_{amb})/5{\cdot}5$, $\therefore$ $T_{amb} = 51{\cdot}5$°C which seems reasonably safe.

If we take the air temperature as 25°C maximum, then we can calculate the junction temperature by the same equation: $7 = (T_j - 25)/5{\cdot}5$. $T_j = 63{\cdot}5$°C, which seems reasonably safe.

Note that the manufacturers of power transistors specify safe-operating areas of voltage and current.

Transistor Bias

The bias for a valve simply sets the operating anode current with no input signal. The bias for a transistor must set the no-signal collector current, but must also prevent thermal runaway. Dissipation in a transistor, and increases in the temperature of the surrounding (ambient) air, cause an increase in the leakage current which flows across the reverse-biased junction between collector and base. This current behaves as an extra bias, being shared between the base circuit and the base to emitter junction, and causes higher current to flow in the collector circuit. In an incorrectly designed circuit, this can cause so much extra dissipation (hence more leakage, etc.) that the collector junction burns out.

This condition is called thermal runaway and is of most concern in the design of output stages using germanium transistors. Where the collector load is a resistor whose value is high enough to prevent excessive current flowing in the transistor, thermal runaway cannot do any damage, though it can cause the collector voltage to 'bottom' (reach to nearly emitter voltage) and so cause distortion or no signal output.

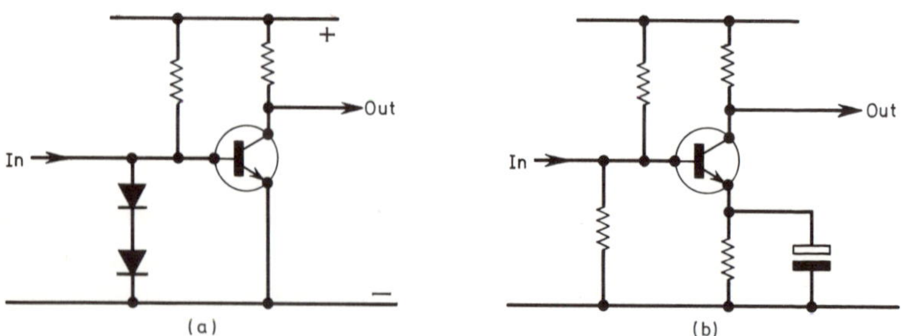

Fig. 9.6 Biasing to reduce effects of temperature change.

The minimum collector load resistance for this purpose is such that with half of the supply volts across it, it dissipates a power equal to the maximum dissipation of the transistor at its maximum temperature. For example, if the transistor is rated at 2W max. at 90°C junction temperature, and the supply is 20V, then 10V across a resistor dissipates 2W with a current of 2/10 = 0·2 amps, or 200mA. The value of this resistor is 10V/0·2A = 50Ω.

When the value of collector load resistor is smaller than this minimum value it is essential to control the bias in such a way that runaway cannot occur under any reasonable conditions of temperature. It is good design practice to bias any circuit so that collector current is fairly constant with temperature even if the collector load is too high to allow damage due to thermal runaway.

Fig. 9.6 illustrates methods of biasing transistors to avoid excessive change of collector current with temperature. A junction diode in the bias circuit may be used so that the change in junction resistance with rising temperature lowers the bias voltage on the base of the transistor, and so lowers the base current. Alternatively, the base voltage may be fixed by a potential divider, and an emitter resistor used so that any unwanted increase in collector current causes the emitter voltage to rise and so bias back the base (a case very similar to the use of cathode bias with a valve). The lower the resistance in the base circuit, the more stable the bias is; the higher the resistance used in the emitter the more stable the bias.

Silicon transistors have much lower leakage currents than germanium transistors, and need much less care with stabilising bias. Germanium power transistors are the most difficult to stabilise, and overheating can often be traced to faults in the bias components.

Transistors used in a.f. amplifiers or in operational amplifiers are often biased in a feedback circuit, which acts to keep the output voltage constant despite changes in individual transistors. Such circuits consist of feedback loops from which a.c. signals are filtered out, sometimes incorporating an extra transistor which is there only for the purpose of amplifying the d.c. feedback signal (Fig. 9.7). As the open loop gain of the amplifier (which must be direct coupled) may be very large at d.c., the stabilisation which is achieved in this way is very great, and this method is now almost universally used in biasing high gain amplifiers.

Checking Bias

Unless stabilisation is 100% perfect, the bias current of a transistor must vary with temperature, and there is little point in checking for some definite bias current, except in power stages. Transistors used as voltage amplifiers should have the collector voltage checked —this voltage should always be enough to allow the whole of the peak-to-peak output voltage to be handled; if this is so, at all working temperatures, biasing is satisfactory. For example, if the supply voltage is 20V, the voltage at the emitter is 2V, and the output voltage is 2V peak, then a bias voltage on the collector of anything between 4V and 18V (2 + 2 to 20 − 2) would be reasonable at the extremes of temperature, though at normal working temperatures the bias voltage should be nearer midway, 11V.

For a power transistor, collector current (sometimes measurable as emitter voltage if an emitter bias resistor is used) must be measured and should be within the limits set by the designer at stated temperatures. Two useful tools for thermal bias checking are a thermometer with aluminium foil wrapped round the bulb (so that it can be 'moulded' round the transistor case) and a hair drier, for heating transistors.

Where the bias is achieved by a feedback network of the type mentioned earlier, bias faults can be bafflingly difficult to solve, for each component in the feedback loop contributes to the bias. The first step is to check voltages all round the loop, looking for a transistor which is cut off or bottomed. If all transistors are conducting normally, but the output voltage is abnormal then a change of resistance in the feedback loop itself is indicated.

Frequency of Operation

The operating frequency of a valve is set mainly by the circuits round it and their time constants, along with the stray capacitances of valve and valve holder. Only at frequencies above 500MHz do we start to encounter trouble due to the time taken by an electron to travel from cathode to anode. The electrons and holes in transistors travel very much more slowly, however, and each transistor has a maximum frequency of operation as an amplifier which cannot be improved by better circuits or by cutting stray capacitances.

The symbol fhfb (or sometimes, hf_{21}) is the frequency at which a transistor's amplification is 3dB down on the 1kHz value when the signal is fed in at the emitter and taken out at the collector ('common base connection'). This is always a considerably higher frequency than fhfe, the frequency at which amplification is 3dB down on the 1kHz value using the common emitter circuit. The maximum frequency at which a transistor can oscillate is higher than either of these values and is sometimes quoted as f_{max}(osc.).

To avoid the need for quoting a large number of different values

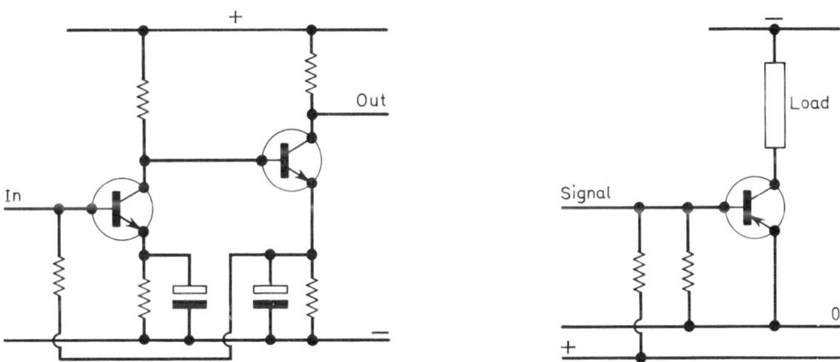

Fig. 9.7 (left) Using a negative feedback loop to stabilise the bias of two stages. The capacitor between the emitter of the second transistor and earth prevents signal voltage from being fed back, except at very low frequencies.

Fig. 9.8 (right) Switching a transistor. In this case, the bias has switched the transistor off, and the signal will switch the transistor *on*. If the bias line were of opposite polarity, the transistor would be biased *on*, and the signal would be a positive-going signal (p-n-p transistor) to switch it *off*.

for the maximum frequency of operation, the figure fr is frequently quoted in modern data sheets. This is (approx.) the value of hfe × frequency at high frequencies (and has no meaning at frequencies where hfe is equal to its 1kHz value) and can be used in very much the same way as the G–B constant of a valve.

The value of fr increases with increasing collector voltage and current, but is a constant for a transistor once the bias is set. At the fr (meaning transition frequency) there is no useful gain, but oscillation is just theoretically possible. At half of fr, current gain of 2 should be possible, etc. It does not follow that at 1/200 of fr, a gain of 200 is possible, because the 1kHz value of hfe may only be 100.

Switching with Transistors

When the base of a transistor connected as in Fig. 9.8 is reverse biased (in this case, to a positive voltage), the current in the collector is only the leakage current, which may be a matter of microamps. When the base is heavily forward biased (negatively in the example), the collector will pass a large current limited only by the voltage of the supply and the impedance of the load, and the voltage between collector and emitter is very low (about 0·2V in typical germanium p–n–p switching transistors).

In each state, the wattage dissipated in the transistor is very low, since either the current is very low or the voltage is; high dissipation occurs only while the transistor is changing states (being highest when both voltage and current are half of their maximum values) and the time which this takes can be very short. The actual wattage dissipated is calculated by

$$\text{wattage during switching} \times \frac{\text{switching time}}{\text{time at rest}}$$

For example, if 20W would be dissipated during switching, the switching time is 1μS, and switching is carried out every 500μS, the effective wattage is $20 \times \frac{1}{500}\text{W} = \frac{1}{25}\text{W} = 40\text{mW}$.

In this way, a transistor with a fairly low dissipation can switch considerable voltage and current loads, because it is the average power which cause thermal damage. There is a limit to what can be done, however, because there is also a limit to the current which can be passed through any transistor, and to the voltage which can be applied across it. So, even if the average wattage is within limits, the switching circuit must be designed so as not to exceed other limits. Note that the transistor must dissipate a high wattage ($\frac{1}{2}$ volts $\times$ $\frac{1}{2}$ current) while in mid-switch so that a slow-switching transistor has to dissipate more power.

Transistor Faults

It is a good general rule that, when a transistor circuit is faulty, transistor failure is the least likely cause unless the circuit is working the transistor to its limits. Most transistors circuit faults are due to failure of bias.

Transistors do not develop low emission, and failure is usually total and irreversible, in the form of s/c or o/c between two terminals. This may be caused by overheating, by excess voltage or current flowing for some time, or by very short spikes of voltage (transients) well above the voltage limit of the transistor. Such transients may be caused by switching, by other circuits working from the same power supply, or by the power supply itself, or they may be picked up from other equipment.

Designers usually use integrating circuits in supply lines and, where possible, in signal lines, to prevent damaging transients reaching the transistors most likely to be damaged—those with low collector loads, such as power transistors. Where failure occurs, and extensive oscilloscope testing shows transients present on supply line or in signal path, a CR integrating circuit must be used to cut down the rate of rise of voltage and limit the amount. Avalanche diodes and similar devices may also be helpful (see later notes).

Faulty bias may cause apparent failure. If the base of a transistor is forced to pass excessive current, the collector voltage falls to the level of the base voltage, and only signals whose peaks cut off the base current stand much chance of getting through. In this case, the bias components should be checked, as they should also be when the base is reverse biased. This over or reverse biasing may be done deliberately in a pulse amplifier and the checks should be made with the normal signal in.

Overheating is the most common source of genuine transistor failure. Hot components nearby (a resistor overloaded, a faulty choke), a soldering iron hooked over a circuit board in the course of servicing, or a hot airstream from a heater can all cause failure. In some cases, perfectly sound transistors are taken out because of bias failure and replaced carelessly with overheating of the new transistor, so that the replacement transistor fails before it is used. When any soldering or unsoldering job is carried out on a printed circuit board which contains transistors, the transistors must be protected against heat. This may require four 'hands', but can cause less trouble later.

'Heat shunting' means holding the lead of the transistor most likely to get hot (base, if that lead is being unsoldered, or if the bias resistors are being removed, etc.) until all the metal around is cool again. Although silicon transistors will stand temperatures higher than that of boiling water it is a good practice not to subject transistors to temperature which the fingers cannot stand.

Remember that the copper used on printed circuit boards is an excellent conductor of heat, and that a soldering iron used to remove a coupling capacitor can have its heat conducted down the copper strip to a transistor further down the board. Where it would be impossible to hold a large number of leads, it can be helpful to pack up parts of the board with aluminium foil to help in conducting away heat which, like electricity, prefers to flow through low resistance paths.

Field Effect Transistors

The first field effect transistor appeared only a short time after the first bipolar (normal) transistor, but development of FETs has been much less rapid. FETs use no base region with minority carriers moving through it, and so are reversible and make ideal switches. The FET basic principle is shown in Fig. 9.9.

Imagine a layer of n-type material which has been 'pinched' to a narrow filament by a p-type junction. The resistance between the two portions of the n-type material to which contact is made depends on the availability of carriers in the pinched region, and this in turn depends on the state of the junction. If the junction is reverse biased (more negative than any region of the n-type material) there is a withdrawal of carriers from the junction region as described earlier for the bipolar transistor and the pinch region cuts off completely, becoming non-conducting.

This type of junction FET may be p-channel or n-channel but is usually found as a silicon n-channel type. The leads to the channel are designated as *source* and *drain* (corresponding to emitter and collector) and the junction lead is the *gate*. The characteristics which are of particular interest are the zero bias current I_{DSS} and the pinch-off voltage $V_{p(GS)}$; these graphs are generally of the shape shown in Fig. 9.10 for each type of junction FET.

These characteristics determine the bias arrangements for the FET, and the amplification is determined by the mutual conductance Y_{FS} (for grounded source circuits), which is of the order of 2–8 mmho. The input resistance is very high, since it is a connection into a reverse-biased junction, and the output resistance is of the order of 20kΩ. Note that if a forward voltage is applied to the gate, current will flow and the input resistance will then be low.

In addition to junction FETs, insulated gate FETs are also found. In this type of FET, the pinch effect is achieved without a

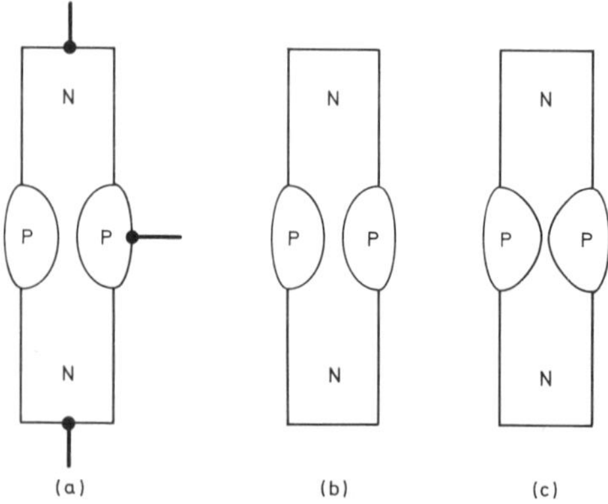

Fig. 9.9 Operating principle of FETs. (a) the p-region is formed so that it pinches the n-channel down to a slender thread; (b) with forward bias (enhancement) or low negative bias on the p-region, current flows normally in the n-channel; (c) with the p-region reverse biased heavily, the n-channel is 'pinched-off' so that practically no current flows.

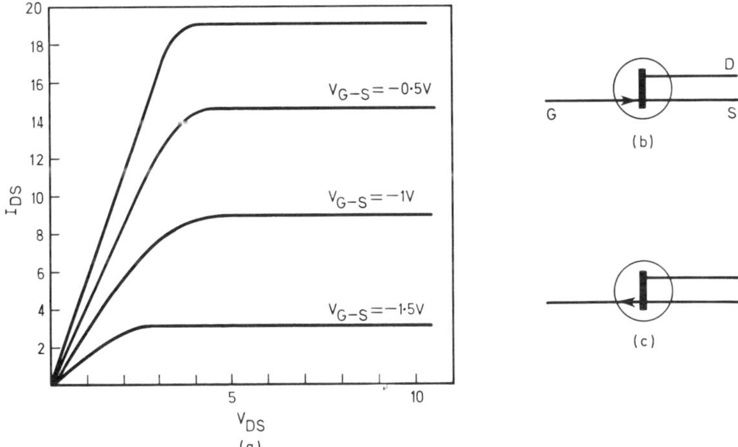

Fig. 9.10 Typical junction FET characteristics. (a) note the similarity to the characteristic of the pentode valve, (b) symbol for n-channel junction FET, (c) symbol for p-channel junction FET.

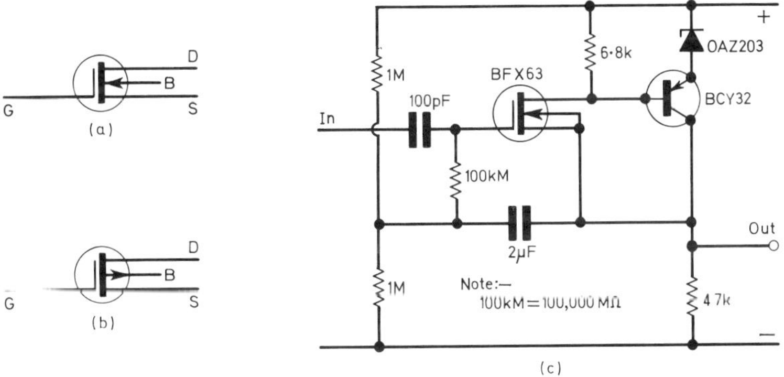

Fig. 9.11 (a) symbol for n-channel MOSFET, (b) symbol for p-channel MOSFET, (c) circuit of impedance converter using an early MOSFET. Performance: Input resistance $10^{12}\Omega$ (1000GΩ); Output resistance 50Ω; Voltage gain 0·97.

junction, making the channel one plate of a capacitor and the gate the other plate (Fig. 9.11). A charge on the gate can then 'hold' a charge on the channel, preventing or encouraging movement. As the gate is, in this design, completely insulated from the channel, the input resistance is very high, as high as the leakage resistance of a capacitor, of the order of 1,000MΩ. The construction usually followed is that a layer of silicon oxide is formed on the silicon channel and a metal layer evaporated on top of the oxide to form the gate.

Such a device is known as a MOSFET (Metal-Oxide-Semiconductor FET) and the symbols used for various types of MOSFET are shown. As the gate is completely insulated from the channel, no current flows whether the gate is positive to the channel or negative, and the MOSFET can be used in the 'depletion mode', where the voltage on the gate ranges between pinch-off voltage and source voltage, or the enhancement mode where the gate voltage assists conduction and ranges between source voltage and the maximum permitted forward bias.

It is normal also with MOSFETs to take out a 'substrate' connection, an additional connection to the channel close to the pinch region, separately, though this terminal is practically always connected to the source when in use. In some MOSFETs, the substrate and the source may be internally connected. MOSFETs have mutual conductance values of about 10mmho (typical) and extremely low drain-to-gate capacitances measured in fF (10^{-15}F). Typical pinch-off voltages are in the 5V to 10V region, with gate leakage currents of 1nA (10^{-9}A) maximum.

Several gates may be formed on one MOSFET, making this a construction of interest in logic design. For amplifiers of extremely high input impedances, MOSFETs are rather superior to the electrometer valves which used to be used, and have the advantage of low-voltage power supplies. The same very high input impedance can be the cause of damage to the MOSFET.

Electrostatic charge built up on the hands or on unearthed materials can cause the gate of a MOSFET to attain high voltages during handling, often sufficient to break down the insulation. To avoid this, MOSFETs are supplied with a conducting rubber ring shorting the gate lead to the drain, and this should not be removed until the MOSFET is safely soldered in circuit with some resistance between gate and drain. When a MOSFET is removed from circuit, some arrangement must be made to keep the gate shorted to the drain. A piece of damp cotton thread is enough; its leakage resistance is less than the input resistance of the gate.

Electrically, the characteristics of the MOSFET are remarkably similar to those of the pentode valve.

CHAPTER TEN

SEMICONDUCTORS II

ZENER DIODES ARE USED FOR VOLTAGE REGULATION. They have a normal diode characteristic in the forward direction, but when reverse bias is applied, current is very low until a certain value of reverse voltage (the Zener voltage) is reached, when the current increases very greatly. At the Zener voltage, a large change of current through the diode causes only a small change of voltage, expressed by the quantity called the dynamic resistance (measured in ohms). For example, a dynamic resistance of 20Ω means that a change of current of 10mA would cause a voltage change of 200mV across the diode (using $V = R \times I$). Note that the dynamic resistance has no relation to the d.c. resistance, which would be the ratio of stable voltage to current flowing and which is of no interest, as it is not a constant. See Fig. 10.1 for characteristics.

Zener diodes are available in various dissipation ranges, with a selection of stable voltages which follow the preferred value ranges

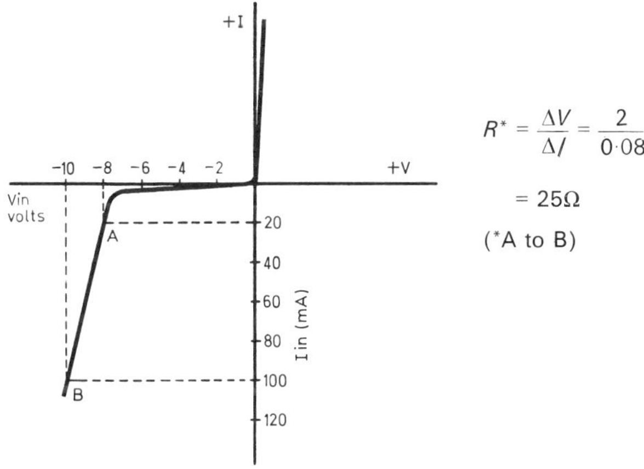

$$R^* = \frac{\Delta V}{\Delta I} = \frac{2}{0{\cdot}08}$$

$$= 25\Omega$$

(*A to B)

Fig. 10.1 Characteristics of a Zener diode, and calculation of dynamic resistance.

outlined in Chapter 2. Temperature coefficients range from very small and negative to fairly large and positive, from about −0·2mV/°C to + 30mV/°C, depending on the dissipation and the stabilised voltage of the diode. Generally speaking, Zener diodes giving a stabilised voltage of about 5·6V have the lowest temperature coefficients, and those with a stabilised voltage between 6·8V and 7·5V have the lowest dynamic resistance. Dynamic resistances of 2Ω or less are common with the high wattage types. Manufacturers indicate maximum temperature ranges, generally −60°C to +150°C, and the derating of dissipation which is required at high temperatures.

Other Diodes

Shockley diodes are used for generating pulses. At low voltages, forward or backward, the diode does not conduct, but at some high voltage, typically 50–100V, it conducts in the forward direction and can pass large currents; in this respect it behaves rather like a gas-filled diode. Fig. 10.2 shows this type of diode used as a pulse generator in a circuit where a voltage charges a capacitor until the Shockley diode conducts and sends a large pulse of current through the load resistor. The output of this circuit is one pulse every cycle, or two pulses per cycle if the switch is closed.

Mixer diodes are used at radar frequencies (1GHz or more) to obtain an i.f. from the received signal and a locally-generated oscillator. They are designed to fit into a waveguide and to transfer the i.f. signal through a coaxial cable.

Avalanche diodes are designed so that, above a certain voltage applied in the reverse direction, they break down and pass a large

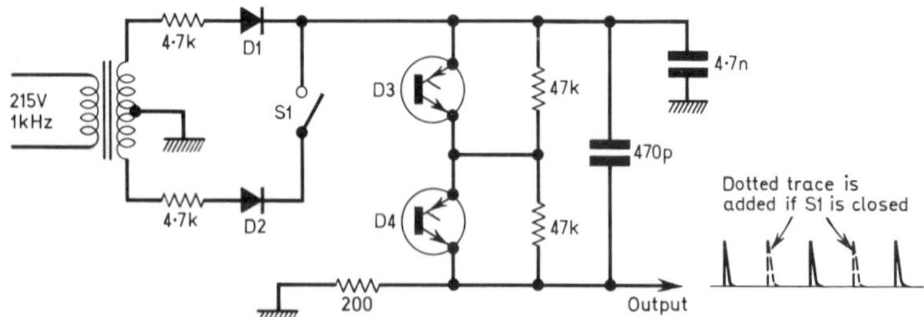

Fig. 10.2 Use of Shockley diodes for pulse generation. The rectified output from D1 and D2 (if S1 is closed) charges the capacitors until D3 and D4 conduct, sending a large pulse of current through the 200Ω load. The diodes remain conducting with a small current, limited by the 4·7kΩ in the power supply, until the voltage supplied by the transformer reaches zero. (*Courtesy Decca Radar Ltd.*)

current without damage. This is basically similar to the action of Zener diodes, but avalanche diodes are used as rectifiers and the carefully controlled avalanche characteristic enables them to remove transients or other surges from the supply.

Tunnel diodes are diodes with a normal reverse characteristic but whose forward characteristic has a peak and a valley (see Fig. 10.3). Between the peak and the valley regions, the voltage across the diode *decreases* as the current *increases*, so that the value of dynamic resistance in this region is negative. The diode cannot remain in this negative resistance region, and will switch very rapidly to another stable state with the same current or voltage, whichever can be changed.

If the diode is connected to a resonant circuit of any kind, it can cause continuous oscillation, and is also useful for rapid switching at low voltages and currents. Typically, a tunnel diode might have a peak current of 1mA, peak voltage of 60mV, valley current of 0·1mA and valley voltage of 350mV.

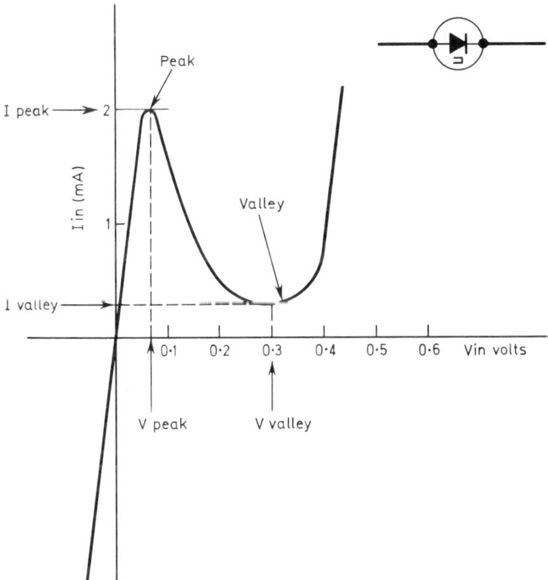

Fig. 10.3 Tunnel diode symbol and characteristic.

Electroluminescent diodes are made of gallium arsenide or indium phosphide in crystalline form, and their behaviour is electrically similar to that of any ordinary diode. When forward current passes, however, the energy dissipated appears in the form of infra-red or visible light radiated from the junction. The light output may be modulated by modulating the diode current, so making it very easy to construct short distance communication linkages. The beam can be made substantially parallel, similar to that from a laser, and it can also be made coherent (remaining in phase) like the output of the higher power lasers.

Gunn diodes, similar in action, provide microwave energy rather than light when current is passed and when a suitable resonant cavity is coupled to them. They have rapidly replaced klystrons as generators of low-power microwave energy.

Thyristors

The thyristor is a four-layer semiconductor, meaning that instead of the n–p–n structure of the transistor, they use a p–n–p–n type of structure with one extra junction. Only three leads are brought out of the five possible connection points, however, and these are termed the anode, the cathode and the gate. If the anode and cathode are connected as a normal diode, with a positive voltage on the anode and a negative voltage on the cathode, no current flows between these elements until current flows in the gate-cathode circuit.

Once triggered in this way, the thyristor conducts and continues to conduct until the current between anode and cathode falls below the minimum 'holding current', or until the anode becomes negative to the cathode. The name (formerly it was called a Silicon Controlled Rectifier, or SCR) is derived from that of the gas thyratron, for which the thyristor is a replacement in its circuit applications.

Apart from the generation of high-power pulses, where the thyristor has obvious applications, the main applications of the thyristor have been in a.c. power handling circuits, such as for controllers of motor speed, lamp brilliance, etc. Two types of power control circuit are in use; phase control (Fig. 10.4b), where the thyristor can be switched on at different times during a positive half-cycle of a.c., and integral cycle control (Fig. 10.4c), where the thyristor is switched on for a certain number of complete half-cycles in each second. Note that one thyristor, being basically a half-wave rectifier, can handle only half a cycle of a.c., and two thyristors (or a device incorporating two thyristors) must be used for full-wave control.

The phase-control method is used for low-wattage lamp loads or for small motors, but creates undesirable surges of current in the mains supply, causing radio frequency interference and even, in

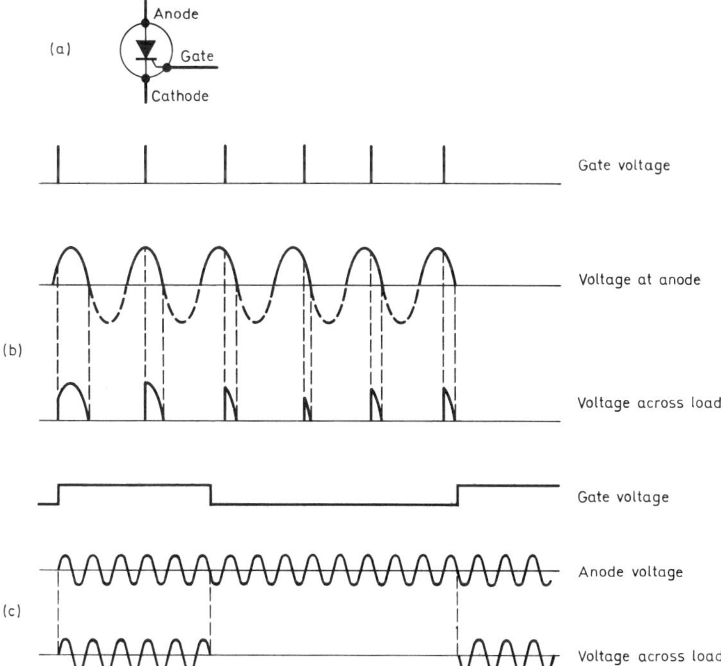

Fig. 10.4 (a) thyristor symbol, (b) phase control—the point during the half-cycle at which the gate fires can be adjusted, (c) integral cycle control—several cycles are allowed to pass at a time, the load averaging out the effect. The triggering of the gate need not be continuous, as shown, but can be for a few milliseconds at each zero-voltage point.

extreme cases, thyristor controllers to interfere with each other. The second method is 'cleaner' as each thyristor is switched on when the voltage across it is zero, so causing no sudden surges of current, but the method can be used only where the load has a long time constant (a motor with a heavy flywheel, or a heater) as the time between power-on and power-off is much longer.

Circuits and data for the use of thyristors in control circuits are available from the manufacturers, and one is shown in Fig. 10.5, but the characteristics need some interpretation. Normally the peak forward volts, with the thyristor non-conducting, equal the peak reverse volts, but the emphasis is on the word 'peak', remembering that 250V a.c. has a peak voltage of nearly 360V and a peak-to-peak of about 720V.

As well as the voltage limits, there are limits on the rate at which a voltage can be permitted to rise when applied between anode and cathode, as a fast rising voltage can cause triggering with no gate current applied. A typical maximum rate-of-rise is 10V/μS; a rate which can be easily attained when a mains supply is switched on or when another thyristor in the circuit switches on. To prevent the rate of rise across the thyristor becoming excessive, a series resistor-capacitor combination may be connected between anode and cathode.

The maximum gate-triggering current and voltage must also be taken into account. It is very easy to design the circuit or to modify it in service so that the voltage or the current is obtainable, but not both together, or to obtain a voltage and current that are marginal, so that the thyristor sometimes fails to fire. Also quoted will be the turn-on time during which the minimum value of gate current must flow to switch the thyristor completely into conduction, and the turn-off time for which the anode-cathode current must be below the holding value for the thyristor to be turned off. Typical values of these two parameters are 2μS and 100μS respectively, though rapid-turnoff thyristors are now obtainable for use in high power, high frequency oscillators.

Large numbers of thyristors are available on the 'surplus' market for which the maximum forward voltage and rate of rise are not closely specified, and these often fail to operate correctly in published circuits. This can be particularly undesirable in car ignition circuits, where a sudden failure of thyristor action during overtaking (often due to rate-of-rise triggering) can have fatal results.

Finally, the peak voltage across the thyristor when it is operational must be known so that the watts dissipated across it can be calculated. This calculation is straightforward in power control circuits but complex in some pulse circuits where large currents flow for short times. To aid calculations, most manufacturers issue safe operating area charts (as for power transistors) showing clearly which limits of voltage and current must not be exceeded for reliable operation.

Triacs (trademark of General Electric of USA) are two-way thyristors in which conduction can be controlled in each direction. The applications and circuitry for Triacs are essentially similar to those of thyristors, with the bonus that one Triac can replace two thyristors with less than twice the amount of other circuit components.

Bilateral switches are a diode version of the Triac (the General Electric version is called a *Diac*) with no triggering electrode. A voltage applied in either direction causes no current to flow until the triggering voltage is reached, when current flows and the voltage across the diode becomes very low. Bilateral switches are used mainly in the gate circuits of thyristors and Triacs to ensure sharp triggering of the gate circuit.

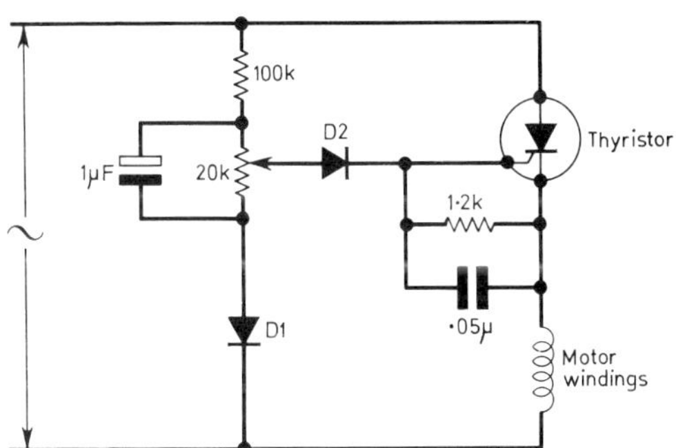

Fig. 10.5 Example of motor speed control circuit using thyristor.

Unijunction Transistors

These have two bases and an emitter, and pass no current from emitter to one base until the emitter has reached a definite voltage, which, for a given type of unijunction, is a definite fraction of the voltage between the bases. This fraction is called the 'intrinsic standoff ratio', and is usually between 0·5 and 0·7.

Once the voltage applied to the emitter is greater than the firing voltage, the emitter current increases due to a form of internal feedback until it is limited only by the resistances of the circuit. The unijunction is also widely used for thyristor triggering, as the firing point can be varied by varying the voltage applied between the bases.

Integrated Circuits

Integrated circuits (ICs) consist of silicon chips on which large numbers of transistors, diodes, resistors and (less often) capacitors have been formed by the same techniques of photo-etching and vacuum evaporation of material as are used for transistors, but which are interconnected on each chip so that the resulting active component carries out a much more complicated circuit action than that of one transistor. Such circuits differ greatly from conventional circuits set up to do roughly the same circuit duty, and some of the notable differences are of considerable importance in the design and servicing of equipment using such circuits.

The size of the complete circuit is the most obvious difference. The silicon chip on which the circuit is formed is the same size as that which would be used for forming a transistor. Because a greater number of lead-in and lead-out wires have to be taken to the chip, however, the final mounted circuit is generally on a base larger than that of a transistor, sometimes of the familiar 'top-hat' form, but now more usually as a 'flat-pack' with seven leads on each side.

As these leads are generally shorter and more robust than those of a transistor there is a de-soldering problem if ICs have to be removed from printed boards, and some form of de-soldering gun, working on the principle of melting the solder and pumping it away, is needed.

Because these leads are connected to the chip by fine gold wires, compression welded, there is only a small flow of heat from the flatpack leads to the silicon chip, and little risk of damage during soldering or de-soldering compared with a germanium (or even silicon) transistor. The reliability of the IC in service is critically dependent on the quality of the welding of these wires, and this operation is very carefully controlled by the manufacturers.

The size of the IC is closely connected with the thermal effects in the device. Any change of external temperature which affects the IC must equally affect all stages incorporated. For this reason, balanced amplifier stages can be designed in which the balance is much less affected by temperature change than could be achieved with separate (discrete) components. Heat generated within the IC, however, affects all of its components, small signal stages as well as large signal stages, and this limits the power dissipation of ICs as the heat cannot readily be removed from the inside of the chip. Careful design has enabled powers of several watts to be dissipated by keeping the high-power stages to the outsides of the chips and attaching heat-sinks.

The physical size of the IC also affects the degree of isolation or interaction between stages (Fig. 10.6). In a circuit mounted on a PCB, it is comparatively easy to reduce stray capacitance or leakage resistance between stages by separating them, designing the board so that critical stages do not come close to each other, and by decoupling all supplies to them. This cannot be done in the design of an IC, though some degree of isolation can be achieved. The whole IC is formed on a chip of silicon whose resistance is high but not infinitely high, and all parts of the circuit will have some leakage resistance to each other.

The silicon also has a quite high permittivity, so that stray capacitance is considerably higher between any two points (not connected) than it would be on a conventional circuit. Because of this, some circuits which work perfectly well in discrete form cannot be made to work if directly translated into the form of an IC, and a different approach to circuit design must be used. From the designer's point of view, it is rather like being required to design printed circuits which will work when immersed in water.

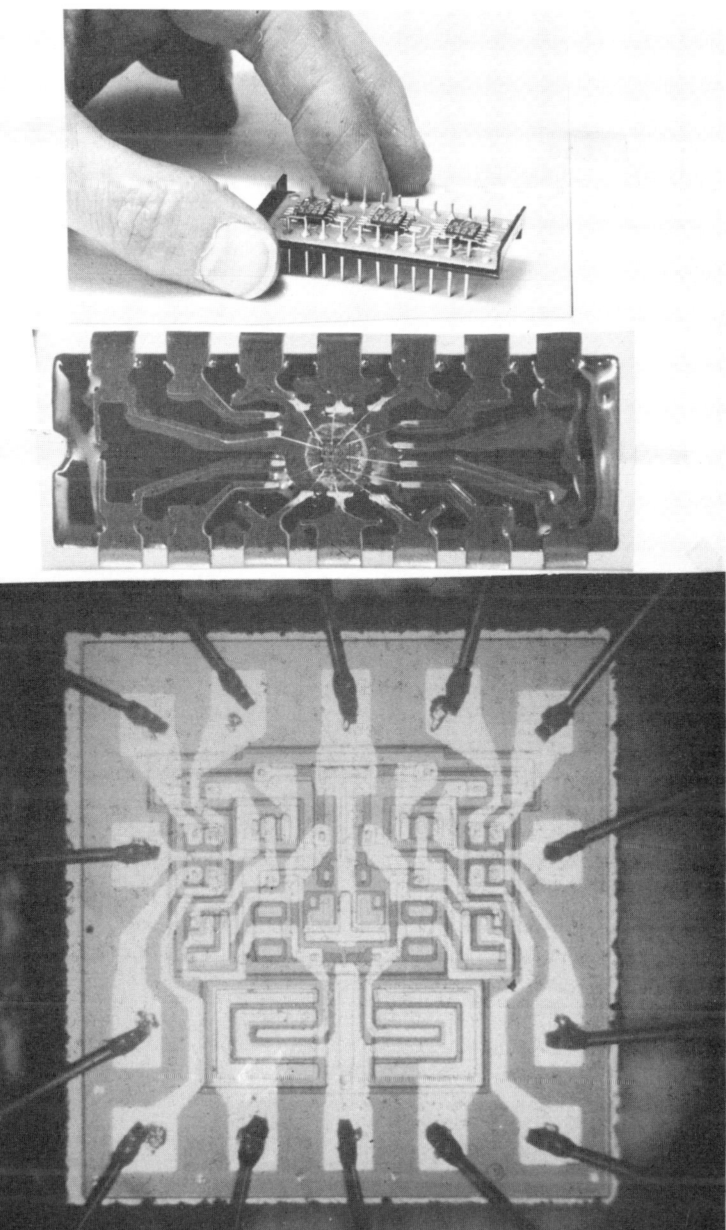

Fig. 10.6 Three photographs of the same IC, showing the scale of the circuit itself. (*Courtesy Elliott-Automation division of GEC*).

The formation of the active devices, transistors and diodes, is particularly easy in IC construction, but resistors are much more difficult to form and have very wide tolerances, and capacitors are extremely difficult. These facts dictate the design of ICs, with the result that circuits using large numbers of active devices in place of resistors or capacitors are favoured: for example, a transistor will be used as the collector load of another transistor if possible. This approach often makes the 'reading' of an IC diagram rather complicated, as it is difficult to distinguish which active components are working as amplifiers and which as constant current loads, etc.

In general, if close tolerance resistors or capacitors are required in a circuit they have to be added as discrete components wired between two or more pins of the flatpack. In addition, because of the use of so many active devices, the gain of an IC amplifier cannot be subject to a close tolerance, since the manufacturing tolerances which determine the gain of each stage affect the gain of each stage, so that the overall tolerance of gain is the tolerance of each stage multiplied by the number of gain stages. The normal technique of designing around ICs must allow for this. Digital circuits present few problems, but linear circuits need enough feedback to ensure that the inevitable differences between ICs do not affect overall gain.

Generally, IC designs use balanced stages with components such as inductors and close tolerance capacitors and resistors added externally, large numbers of active devices, and circuits devised to overcome the interaction between stages.

Digital Circuits

Digital ICs are pulse circuits used for counting and controlling operations whose inputs and outputs consist of only two voltage or current levels, one coded as 1 and the other as 0. The user is usually free to decide which voltage level (high or low) is to be taken as the '1'. No other voltage level is of interest, and all of the operations of the circuits are carried out by changing these levels at various points in the circuit.

Counting and controlling operations are carried out by using figures represented in the Binary Scale, where the digits 1 and 0 only are used. This means that digital circuits are not greatly affected by tolerances, by noise in signals, by variations in amplification or any of the other disturbances which affect linear circuits, though they cannot ignore their effects (See Table 10.1).

Digital circuits are not, of course, new but their use has enormously expanded with the arrival of ICs, which now make the

Decimal	Binary
1	1
2	10
3	11
4	100
5	101
6	110
7	111
8	1000
9	1001
10	1010
11	1011
12	1100
13	1101
14	1110
15	1111
16	10000

Addition: 12 + 2

```
 1100
   10
 ----
 1110 = 14
```

Subtraction: 12 − 5

```
 1100
  101
 ----
  111 = 7
```

Multiplication: 5 × 3

```
  101
   11
 ----
  101
 101
 ----
 1111 = 15
```

Division: 12 ÷ 4

```
100)1100(11 = 3
    100
    ---
     100
     100
```

Binary to Decimal:

```
  1   1   0   0   1   0   1   1     Binary numbers
  ↑   ↑   ↑   ↑   ↑   ↑   ↑   ↑
  |   64  |   16  |   4   |   1     Decimal number for
 128     32       8       2         each 1 in column
```

= 128 + 64 + 8 + 2 + 1 = 203

Decimal to Binary: 117

```
                                               0  128
                                               1   64
Subtract nearest power of two: 117–64 =  53
                                          32  →  1   32
                                          --
                                          21
                                          16  →  1   16
                                          --
                                           5     0    8
                                           4  →  1    4
                                          --
Binary = 1110101                                 0    2
                                           1  →  1    1
```

Table 10.1—BINARY SCALE AND ARITHMETIC PROCESSES

assembly of complex digital circuits merely a matter of wiring a few ICs together. The uses and techniques of digital circuitry, both in conventional and in integrated circuitry, are dealt with in the companion to this volume, but some of the terms used to describe digital ICs will be listed here.

Logic level refers to the voltages which are referred to as 0 and 1. There is usually a fairly wide tolerance on these, a typical example being a low level of 0V to 0·4V and a high level of 3·9V and 6·3V.

Threshold voltage is the input voltage at which a device changes over its output or otherwise responds to an input. The threshold voltage may not be the same for a change from high to low as for a change from low to high, and the value of this voltage or voltages decides what tolerances are permissible in other parts of the circuit.

Signal Noise Immunity (SNI) is defined as the voltage difference between the normal signal level (at its voltage nearest to the threshold voltage) and the threshold voltage itself. For a circuit which is at the limit of tolerance on the logic level voltage, the SNI measures what noise voltage would just switch the circuit over.

Fan-out is the maximum number of stages which can be driven from the output of the IC, assuming that the driven stages all take equal currents. Usually it is assumed that the stages being driven take the same current as the stage which is fanned out, and this will practically always be so in a series of compatible ICs. *Fan-in* is the opposite, the number of inputs which can be connected so as to control the output of one IC.

Propagation delay time is a measure of the time taken for a change of voltage at the input to affect the output, and is the average of the times for switch-on and for switch-off, measured in nS.

Apart from these terms above, the words *bit* (sometimes byte) and *clock* occur throughout descriptions of digital circuits. Computing circuits are counting circuits, each simple in itself, arranged in complex forms to count according to a set of rules laid down by a programme of instructions. The counting rate is set and each operation is carried out by a triggering pulse applied to various parts of the circuit from a master pulse generator.

The trigger pulse is known as the clock pulse, since it occurs at regular intervals and controls the timing of all events in the computer. Counting is carried out using binary arithmetic, as already mentioned, where the digits used are 0 and 1. Each digit is a piece of information, a bit (shortened version of *bi*nary dig*it*) which may be stored or passed over to another circuit when the clock pulse arrives.

Before trigger					After trigger	
T	J	K	Q_1	Q_2	Q_1	Q_2
1	0	1	—	—	0	1
1	1	0	—	—	1	0
1	1	1	0	1	1	0
1	1	1	1	0	0	1
1	0	0	0	1	0	1
1	0	0	1	0	1	0

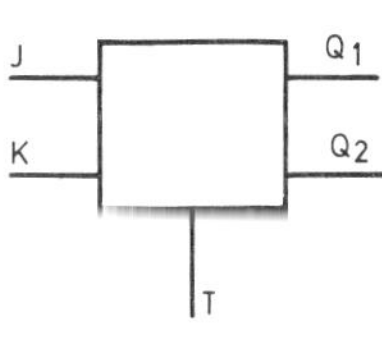

Fig. 10.7 Truth Table for a J-K flip-flop. The Truth Table is a diagram showing what outputs can be expected after the clock (trigger) pulse when various inputs are present.

Not all processes are clock-controlled. Gating, for example, is a logic process—one of deciding what bits shall be counted and which shall be discarded. No clock pulse is needed for these operations, which give an output of constant voltage level, either 0 or 1.

The counting operations are carried out by bistable circuits which perform an operation each time a clock pulse is received. The basic operation of a bistable is division by two, and its output when it is activated by a clock pulse depends on its present input and its last input. When the input to the bistable is a 1, then at the instant of the clock pulse, the output will switch to a 1 or to a 0 depending on whether the previous output was 0 or 1; in other words the bistable changes state if the input is 1 and does not alter if the input is 0. In this way, the number of 1 bits can be added and a carry signal obtained for the next column of addition.

To be of more use, subtraction should also be possible, and this can be carried out if the bistable can be controlled by two inputs. Even more can be done if an opposite-polarity output signal is also available from another point. It is one of the main advantages of integrated circuitry that increasing the complexity of the circuits makes very little difference to the final cost of the circuits, provided they can be used in large numbers. Finally, it is desirable in such counting circuits to be able to set all the counting stages to zero, or to any number which has to be placed in the counter (from a memory store, for example).

In the integrated form, the complete counting stage is usually a J–K flip–flop whose typical truth table is shown in Fig. 10.7. The inputs are applied to the terminals labelled J and K, the outputs are taken from points *Q*1 and *Q*2, and the setting pulses applied to *S*1 and *S*2. The two sets of circuits inside the IC are interconnected.

Integrated circuitry also enables combinations of circuits to be made up. Shift registers can be obtained in integrated form (they include the carry-over which takes place when binary numbers are added) so that a large portion of circuitry can be replaced. Complete (small) memory stores can be obtained in integrated form, suitable for storing and retrieving digital information, and usually with the ability to recover information from any part of the store (random access).

Binary-to-decimal decoders, which give an output in decimal numbers suitable for application to display tubes when their input is a binary coded number with one bit applied at each clock pulse, can be supplied in integrated form, and the once familiar gas tube counters are now being replaced by the faster counting assemblies of adders, shift registers, and decoders driving gas display tubes.

Linear ICs

In a linear circuit, the amplitude of the output is proportional to the amplitude of the input, so that the output voltage plotted on a graph against the input voltage is a straight line. Linear ICs have taken longer to appear than digital ICs because of the tolerance difficulties, and most are balanced amplifiers over their high-gain stages requiring external feedback to stabilise their operation.

Typical of such linear circuits are the *operational amplifiers,* balanced amplifiers of very high gain before feedback, which account for most of the linear ICs manufactured. Also in the linear class come the audio amplifiers, similar in many respects to operational amplifiers but with higher power outputs, and some circuits intended for specialised use within TV sets and which combine linear and digital circuitry.

The main characteristics for a linear IC, taking an operational amplifier as typical, are the nominal supply voltage, voltage gain (open loop) input and output resistances, input offset voltage and current, output voltage swing and operating temperature range. Of these figures, the voltage gain quoted will be a 'typical' figure, with a tolerance of several hundred per cent, and large enough to allow a considerable amount of feedback. Input and output resistances will again be quoted as typical figures, usually with high input resistance of the order of 100kΩ, and low output resistances of a few hundred ohms or less.

Input offset voltage is the difference in voltage between the two inputs of the balanced amplifier for zero difference between the output terminals, or zero output if there is only one output, and the input offset current is the current which flows between the input terminals to maintain this offset voltage. There may be provision in the circuitry surrounding the IC to reduce the offset to zero, though it is normally extremely low.

The output voltage swing will be quoted as ± peak volts for various values of load resistor, if an external load resistor is used. The common mode rejection ratio and power supply rejection ratio may also be quoted; they measure the effective balance of the amplifier by showing how far below (in dB) the maximum output are the outputs from signals applied simultaneously to both inputs or to the power line respectively.

Inputs and Outputs (Digital)

As there is no access to the 'components' within an integrated circuit, the inputs and outputs are of particular significance. In most cases, inputs will be to the biased base of an n–p–n transistor, and outputs from a collector or from an emitter-follower circuit, but the precise characteristics will depend on the type of circuit used.

In a digital IC, the manufacturer will specify the two values of voltage which must be applied at the input to switch over in each direction. These voltages may be quoted for SNI = 0, meaning that the voltage quoted is the absolute minimum, with no allowance for noise; they will also be quoted for other values of SNI to show how tolerant to the presence of interfering signals the circuit can be. As the input voltage is usually being applied to a semiconductor junction, a maximum positive voltage will be quoted.

For example, the Mullard FCH231 gate circuit quotes an input 'high' voltage of 2·3V absolute minimum (SNI zero) to 3·4V for SNI of 1·1V and an absolute maximum of 6·3V. The 'low' voltage for switchover to the opposite state is given as 0·8V for SNI zero to 0·4V (maximum) for SNI or 0·4V. The maximum currents at the inputs are also quoted.

In the example above, they are 25μA at the high input and 2·0mA at the low input, the reverse of what might be expected because the high input is, in fact, biasing a diode to cutoff, so allowing the bias on a transistor to remain unchanged, and the low input is opposing this bias. Designing round the input of such an IC therefore means that the preceding circuits must be able to supply the currents required at each of these voltage levels. If more than one input is to be driven, the driving stage must be able to supply the total current required; this is the principle of fan-out.

At the output of a digital IC, the high output signal is set usually by the supply voltage, which must be controlled. For a circuit using a large number of ICs this means using a stabilised voltage supply, for a change in current must no cause a change of supply voltage so great as to cause a noise signal which will switch ICs on or off. The low voltage output (maximum) must be stated, and also the current available to drive other stages.

If the ICs are to be compatible, the maximum output current must be available at the voltage state (high or low) where the next input takes its maximum current. If an IC has a current output of 12mA at the low state and 50μA at high, it cannot be successfully connected to one which requires 20μA at low and 2mA at high, unless a buffer stage, consisting of a transistor amplifier, is wired between the ICs. This problem will not arise when one family of ICs from one manufacturer is used, but can be troublesome if, for any reason, different types are mixed.

The output details will include the maximum fanout under various conditions. The maximum fanout depends on how stable the voltage supplies and the temperature range of working can be held, so that several figures will be quoted, often one for stable voltage conditions, one for constant temperature but with the connected circuits at the limits of supply voltage, and one for the worst case where one circuit is at maximum temperature allowed and maximum supply voltage and the other connected circuit is at minimum temperature and supply voltage.

The supply voltage, as a nominal value with tolerance limits will be quoted, also the normal and maximum currents drawn from this supply voltage. Also printed among the design data when relevant will be a 'truth table', which is a summary of the operation of the IC which enables the action of the IC to be checked. A truth

The code consists of three letters followed by three figures, possibly with a final letter. The first two letters indicate the family of devices, using the first letter F or G if the device is digital. The second letters are used in alphabetical order as new families are developed. The third letter is used to show the use of the device. The first two figures are the serial number. The last figure shows the ambient temperature range.

Third letter code
H . . . gates, gate arrays, or similar.
J . . . bistable or multistable circuits, flip-flops, registers, or counters.
K . . . monostable circuit.
L . . . level converter circuit.
Y . . . miscellaneous.

Last figure code
1 . . . 0°C to +70°C or +75°C
2 . . . −55°C to +125°C
3 . . . −40°C to +85°C

Table 10.2—THE FC PRO-ELECTRON CODE

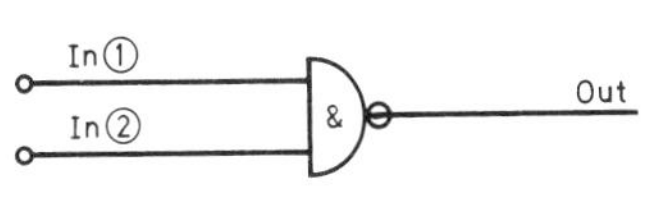

Input①	Input②	Output
Low	Low	High
High	Low	High
Low	High	High
High	High	Low

Fig. 10.8 FCH231 Truth Table.
(*Courtesy Mullard Ltd.*)

table shows what inputs will cause what outputs. For example, for the FCH231 already mentioned, the truth table is shown in Fig. 10.8.

The truth table can be used to interpret what the gate circuit is doing in this case. If we take the high voltage as representing a logic 1 and the low voltage as 0, then the four states read:

$$0 + 0 = 1; 1 + 0 = 1; 0 + 1 = 1; 1 + 1 = 0.$$

This is a NAND gate, meaning that two inputs are needed to change the output from 1 to 0. NAND is a compressed version of Not-AND; an AND gate would have two 1 inputs changing the output from 0 to 1, this gate performs the opposite change, and since the opposite of any action in binary is the NOT-action, the gate becomes a NAND gate.

If we now take the low voltage as 1 and the high as 0, the gate is a NOR gate, and the four states are:

$$1 + 1 = 0; 0 + 1 = 1; 1 + 0 = 1; 0 + 0 = 1.$$

so that with one input OR the other a zero, the output is a 1.

Inputs and Outputs (Linear)

Taking an operational amplifier to be typical of most of the linear ICs encountered, the supply voltage is specified and may be earthed at one side or at a centre-tap. The range of input voltage will be specified; both as common-mode input, the signal applied to both inputs in the same polarity, with very little output; or in the differential mode where one signal is applied to one input and an equal-amplitude inverted signal applied to the other input, causing the amplifier to operate at peak gain. Offset voltage and current have already been mentioned, and the differential input resistance (the resistance measured between the input terminals with the device working) may also be specified.

At the output, the peak current which can be supplied and also the output resistance will be quoted if the final load is 'built-in'; if an external load resistor or impedance is used, then the range of possible values will be quoted along with the peak voltage swing obtainable. The remaining characteristics are then the common mode rejection ratio and the open loop voltage gain.

In addition, the frequency response must be specified. In the earlier designs of operational amplifier, provision is made for connecting external networks to adjust the response, particularly the phase shift at high frequencies. Since operational amplifiers are often called upon to provide (with the aid of feedback) very closely specified gain and phase characteristics, these compensations are important, and are labelled 'lead' and 'lag' according to their effect on phase shift.

The lag correction usually consists of a series CR network from a point close to the final output to earth, and it limits the gain at the higher frequencies. The lead compensation usually consists of a feedback network which also contributes to a more linear graph of gain against frequency. Various values of lead and lag compensating components may be used to attain any desired phase and gain characteristic.

More specialised linear ICs exist, for example deaf-aid amplifiers, radio receiver i.f. and detector circuits, audio amplifiers and pre-amplifiers, colour TV processing circuits; all need very much more information to be supplied than can be dealt with generally.

Voltage regulator circuits consist of a linear differential amplifier whose output is proportional to the voltage difference between a stabilised reference voltage (supplied by a Zener diode) and the output voltage. The current passing to the output is controlled by

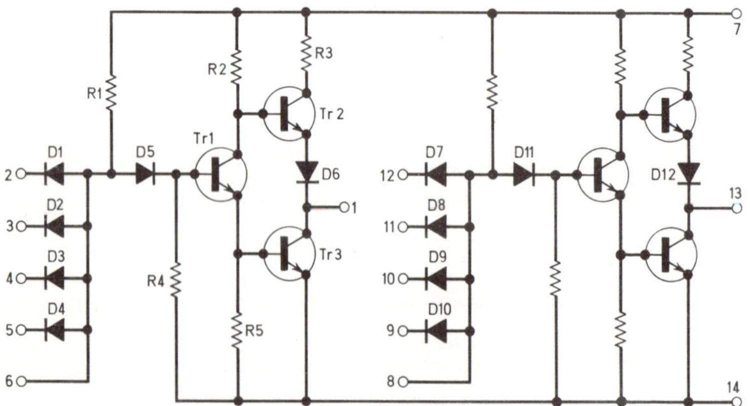

Fig. 10.9 Equivalent circuit of FCH231 (*Courtesy Mullard Ltd.*)

the output of the differential amplifier so that there exists a complete feedback loop acting to keep the output voltage constant. The controlling stage may be a low-power stage built in to the IC, or it may be an externally-wired power transistor if higher powers are to be handled.

The output can be made controllable, by altering the fraction of the output voltage fed to the differential amplifier and the output can be protected against short-circuits by a current detecting stage which switches off the differential amplifier when excessive current flows. The information supplied with such an IC must include the form of the external components needed for control of the voltage, the range of permitted input voltage and the range of permitted output voltage and current.

The FCH231 DTL NAND/NOR Gate

The equivalent circuit of this Mullard IC, taken as a single example of a wide range of digital ICs, is shown in Fig. 10.9, and consists of two separate gate circuits, each consisting of six diodes and three transistors. Inputs are taken to the diode cathodes at pins 2, 3, 4, 5 in one gate and at 9, 10, 11, 12 in the other. The remaining connections (6 and 8) to the input are known as expanders, and are used for altering the bias on the first stage so as to permit more inputs if needed.

The first transistor of each gate is normally biased on through the resistor and diode connecting its base to the positive line, so that, with no inputs, the output at pins 1 and 13 will be low, with Tr2 (taking one gate as an example) switched off and Tr3 fully on. A low input at one or more diodes overcomes this bias and allows the first transistor of the gate to switch off, the second to switch on and the third to switch off. The output of the gate is therefore high if any input is low, and is low only if all of the inputs are high or isolated.

Note that a low input cannot draw current from a high input because of the diodes at the inputs; a low input reverse-biases the remaining diodes. This type of gate is termed DTL because of the use of diodes and transistors (DTL meaning Diode–Transistor Logic). Other ICs exist using TTL (Transistor-Transistor Logic) where the inputs are to several emitter leads formed on one transistor.

The TBA221 Operational Amplifier

The Mullard TBA221 is an operational amplifier with a typical open loop voltage gain of 100,000 and typical input resistance of 1·0MΩ. The equivalent circuit diagram is shown in Fig. 10.10. The differential inputs are applied to pins 2 and 3, and the output voltage is in the same phase as the input to pin 3. The maximum offset voltage between the inputs is 6mV, and this can be reduced to zero by applying the negative supply voltage to the tap of a potentiometer whose ends are connected to the emitters of Tr5 and Tr6 through pins 1 and 5.

Any common mode signal appearing in the same phase on the bases of Tr1 and Tr2 appears at the common collector, whose load is the base–emitter junction of Tr8, to be applied to the base of Tr9. The load of Tr9 is the constant-current transistor Tr10 which has a high dynamic impedance. The base voltage of Tr10 is set by the chain Tr12, R5, Tr11, which also sets the bias on Tr13. The signal at the collector of Tr9 is applied to the bases of Tr3 and Tr4, so completing the negative feedback loop to the emitters of Tr1 and Tr2 for in-phase signals. The typical common mode rejection ratio achieved by this circuitry is 90dB.

Signals applied differentially to the bases of Tr1 and Tr2 appear differentially at the emitters of these transistors, to be current-amplified by the common-base pair, Tr3 and Tr4, a pair of p-n-p transistors. The output signal appears at the collector of Tr4; the transistors Tr5 and Tr6 form a load for the current signals from Tr3 and Tr4, and the differential action is increased by Tr7 which feeds some of the current signal from Tr3 to the bases of Tr5 and Tr6. From the collector of Tr4, the signal is applied to the amplifier triplet composed of Tr16, Tr17 and Tr22. The first two form a high-gain Darlington pair, and Tr22 applies a very large amount of negative feedback to the base of Tr16.

The collectors of Tr16 and Tr17 are loaded by Tr18, whose value of dynamic resistance is controlled by R7 and R8, and the signal at this point is applied to the base of Tr20. A large fraction of the same signal is applied to the base of Tr14, and some is fed back to Tr16 through C1, so that external compensation for frequency response is not needed. This output is then taken to the base of Tr14, which with Tr20 forms a complementary symmetry output stage.

The TBA221 differs from the previous generation of operational amplifiers in the extensive use of complementary symmetry and in the totally internal compensation, and is a good indicator to the trend of design of linear integrated amplifiers. A fourteen pin flatpack version, the TBA221A is also available.

Working with ICs

From the design point of view, ICs have to be treated as a single component, to be biased and fed with inputs with specified source resistances and to be matched to load impedances like any other component. Digital units generally require only a single supply-line and earth, and have few biasing problems; linear units require bias, and may use both positive and negative supply lines balanced about earth. As most linear amplifiers have very high values of gain, a feedback biasing system is essential if the bias has to be applied externally.

From the servicing point of view, ICs cannot be repaired any more than transistors can be repaired; they are used because of their enormous reliability more than for any other reason. If an IC is found faulty, then the entire IC must be replaced, and the surrounding circuitry carefully checked to find a reason for the failure. The easiest way to replace the IC is to snip the leads rather than to unsolder, then to unsolder the tag-ends remaining in the PCB. If circuits are to be tried out, a holder for the ICs should be used; such holders are obtainable for all of the common lead patterns used.

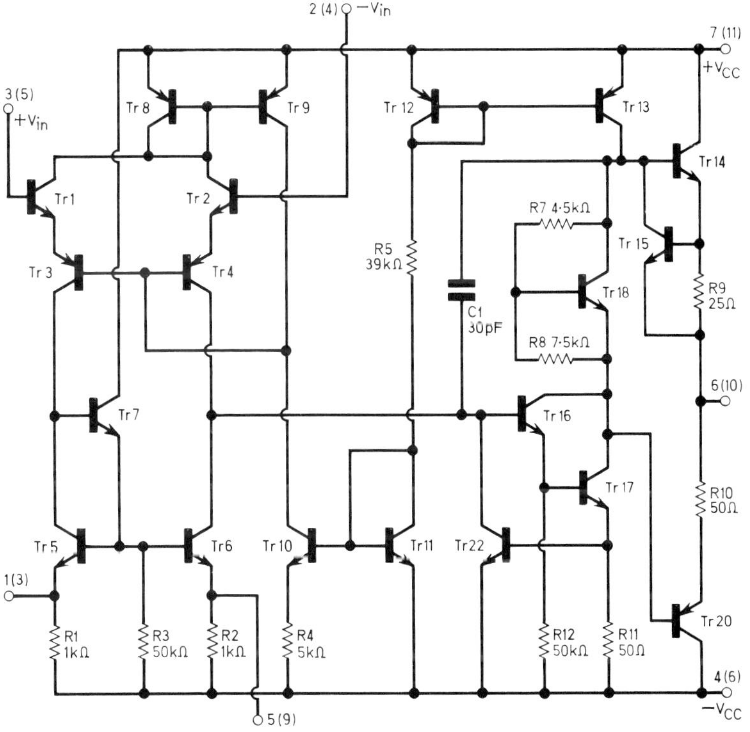

Fig. 10.10 Equivalent circuit of TBA221 (*Courtesy Mullard Ltd.*)

CHAPTER ELEVEN

ELECTROMECHANICAL COMPONENTS

VERY FEW ELECTRONIC MACHINES are entirely electronic; most require at some stage to convert electrical energy into sound, light or mechanical energy. Thus a radio receiver requires a loudspeaker, a television set uses a c.r.t. and most industrial electronic equipment uses electromechanical components such as relays. Included also in this chapter are some assorted devices such as thermal switches which do not readily fit under any other heading.

Relays

An electromechanical relay is a device which performs a switching action when current is fed to an inductor electrically insulated from the switch points. The operating principle is of electromagnetism: a coil of wire with a steady z.f. current flowing in it behaves as a magnet, and is capable of attracting materials such as iron, the magnetism ceasing when the current ceases (Fig. 11.1). The intensity of the magnetic force can be greatly increased if the coil is wound round a core of soft iron. In most relays, the coil is iron-cored and is said to be energised when current passes.

If the energising current is sufficient, an armature moves and so operates the contact which performs the switching action. A large number of contacts may be operated by one coil, but more contacts require more power to be dissipated in the coil. Because of their apparent simplicity designers frequently misuse relays, causing service problems, and we shall examine some of the choices which must be made when a relay is selected.

The operating coil will require a specified minimum wattage (sometimes quoted as operating current or voltage for a coil of stated resistance) to operate the relay. This will depend on the frame size, the coil impedance and the number and size of contacts. There will also be a maximum wattage figure above which the heating effect of the current in the coil may cause damage. This

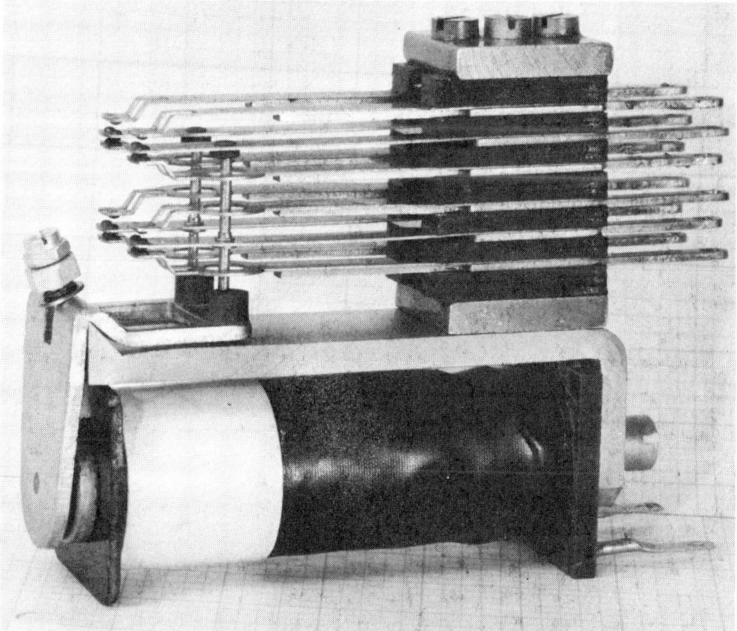

Fig. 11.1 Post Office type relay, clearly showing armature and contacts. This particular type is intended for high voltage switching and is well insulated (*Photo courtesy G.E.C. Telecommunications Group*).

maximum wattage figure will be high if the coil is operated intermittently, low if the coil is operated continuously.

In addition to these ratings, there will be a maximum figure for the voltage between the coil connections and earth. This figure depends on the insulation of the coil and is important if the relay is used to check excessive current in an h.t. circuit, when the coil is in series with the high voltage load. The resistance of the operating coil determines how suitable the relay will be for matching to a valve or a transistor or by any other circuit which will drive it.

The inductance of the operating coil must also be known, for that, along with the speed with which the coil is open-circuited, when the coil is de-energised, enables the designer to calculate what peak volts will be developed across the relay coil. This peak voltage is given by the formula $V = L \times I/t$ (where I = operating current through coil and t = time to switch off) so that, for example, the voltage 'kick' across a 0·1H relay coil which has 100mA flowing and is switched off in 10mS is 0·1H $\times$ 0·1A/0·01S (units: Henries, Amps, Seconds) which equals 1V, hardly a cause for concern.

Imagine, however, a transistor operated relay, inductance 0·1 H carrying 100mA (0·1A). If the transistor is switched off in 100μS, the voltage developed is $0{\cdot}1 \times 0{\cdot}1/0{\cdot}001 = 100$V. This voltage is the same in sign as the voltage at the line end of the relay, so that for an n-p-n transistor (positive collector voltage) this would be a reverse voltage which could damage the collector–base junction. Some of the methods used to avoid damage of this sort are shown in Fig. 11.2.

The contacts of the relay need equally careful consideration. In some applications they may spend most of their life closed with current passing and in other cases they may be mainly open with the contacts exposed to corrosion from the atmosphere. The relay may be switched from open to closed contact intermittently, or it may be opened and closed regularly. An ideal contact material would have very low resistance, would need very little force to keep the contacts together in a low resistance condition, would be robust, resistant to corrosion, long lived, and cheap. There is no such material, and contact materials must be chosen according to the sort of application for which the relay is intended.

Because the contacts have to be moved together mechanically before current can flow in the contacts circuit, there is a lag between applying current to the coil and measuring current through the contacts; this is called the 'operating time'. The operating time depends on the inductance of the coil, the weight of the moving parts, the resistance of the spring which is used to reopen the contacts and the amount of force which can be applied magnetically by the coil.

The operating time is quoted for most relays (sometimes termed the 'make time'); it is seldom less than 20mS, and may be as much as 200mS. The 'break time' is the time from cutting off current to the coil to the opening of the contacts. It depends on the minimum current needed to keep the relay closed, the inductance of the coil, the weight of the moving parts and the spring force; and is also between 20mS and 200mS.

Slugged relays are relays whose coils have been modified by using a copper sleeve so that magnetic force in the coil rises and falls only slowly and make and break times are fairly long.

Uses of Relays

Relays are used where switches would require impossible gangings (where, for example, a large number of contacts must be made, broken or changed at different parts of a large piece of equipment), where switches could not be manually operated, where a sequence of events must take place automatically, where operation of equip-

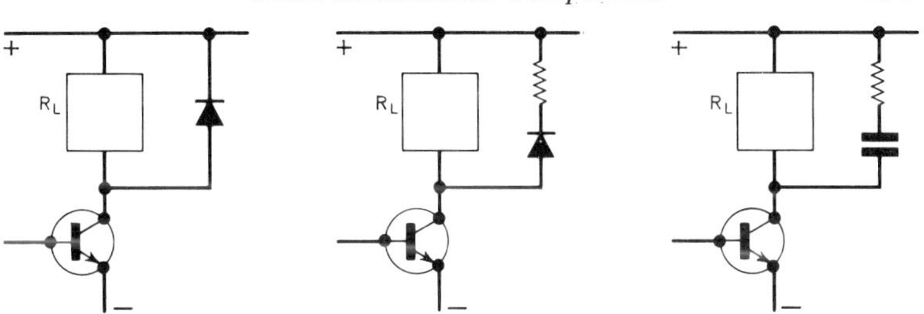

Fig. 11.2 Methods of protecting from the induced voltage when a relay is switched off.

ment requires higher voltages or currents than can be handled by the electronic circuits, or where automatic switching (for example, when excessive current is taken) is required.

Examples include the delayed switch-on of high voltage in a transmitter, overcurrent protection of power packs, range switching in radar installations. Most relays require d.c. voltage but relays can be constructed so that the armature pulls in and holds using a.c. to the coil. A.C. relays tend to be noisy and to go out of adjustment more readily than d.c.-operated relays; for most purposes a suitable supply can be obtained from a.c. by using a simple rectifier circuit.

Relay Diagrams and Circuits

Although a few 'private systems' of showing relays on circuit diagrams are still to be found, most relay diagrams follow the pattern set by British Standard 530: 1948. This standard covers many cases not required in electronics, and only the features most frequently encountered are mentioned in this chapter. The British Standard should be consulted for further information (Fig. 11.3).

The standard system uses the method of detached contacts. The relay coil and each of its contacts are shown at the points where each occurs in the circuit, and not together, as they would be in practice. This means that, for example, where a transistor operates a relay which switches on power to a transformer, the relay coil is shown connected in the collector of the transistor, but the contacts are shown connected in the transformer circuit, which may be at a different part of the diagram. The connection between them is established by giving each a reference number which shows the number of the relay, the total number of contacts and the contact which is indicated by the reference.

The coil of the relay is indicated as a rectangle with a reference

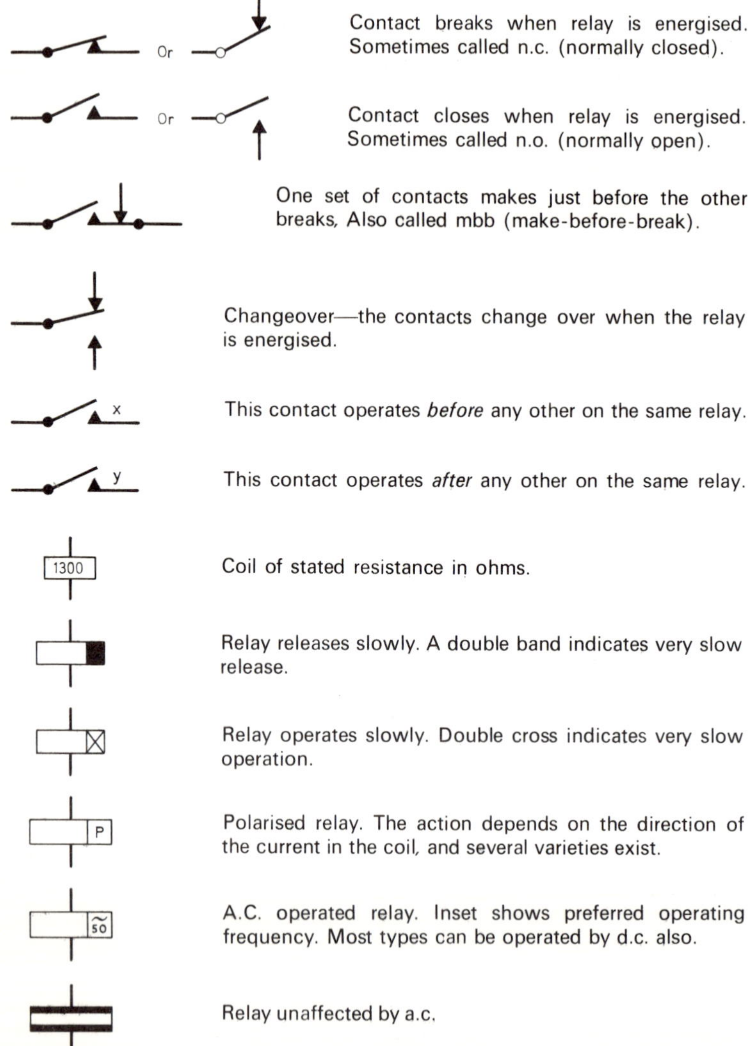

Fig. 11.3 BS 530:1948 Relay diagram symbols (*reproduced by permission of the BSI*).

number written as a fraction, for example R/3. The numerator (top of fraction) is the reference letter or number for the relay (A, B, C, D, etc.) and the denominator (bottom of fraction) is the total number of contacts (one make, one break, or one changeover counting as one contact) used (there may be more available). Each contact is indicated by the same reference letter or number and a number which indicates the number of the contact. Thus R/3 would have contacts R1, R2, R3.

The rectangle used to denote the coil may carry a number inside the rectangle indicating the coil resistance in ohms, or it may indicate in coded form, an unusual operating speed, a polarised relay, an a.c. relay (with the preferred frequency of operation shown) or a relay unaffected by a.c., etc. Most relays used in electronics are straightforward types, and the contact symbols are of more interest. Contacts are shown, as switches, in the position which they take up with the relay unenergised so that an n.o. (normally open) contact closes when current flows in the relay coil, and an n.c. (normally closed) contact opens when current flows. A c/o (changeover) contact takes up a position when energised opposite to that shown on the diagram.

Using relays, a number of sequential switching circuits can be made up at considerably less cost than would be possible by purely electronic means, and one example is shown in Fig. 11.4.

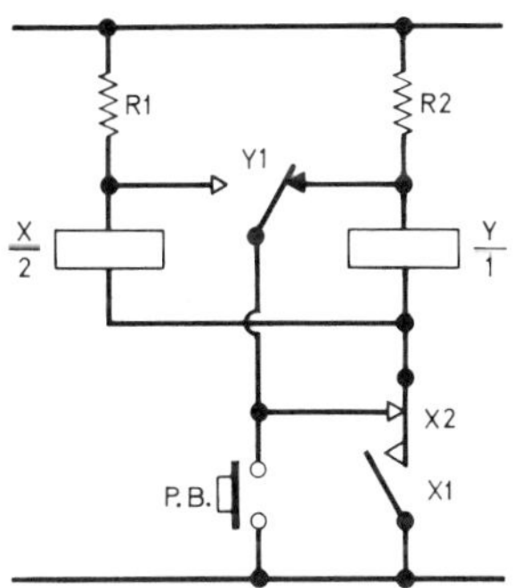

Fig. 11.4 Switching in sequence using relays. When voltage is applied, neither relay can operate because there is no path for current. When the push button is pressed, relay Y is shorted out through contact Y1 and cannot operate. Relay X operates through contact X2 and the push button switch, X1 closes then X2 opens. The current path through X and X1 keeps relay X operating. When the push-button is released, Y operates through X1, since the short circuit path through X2 is now open. Y1 changes over, so that when the push button is closed again, X is shorted, X1 opens and X2 closes, keeping the short circuit. Opening the push button switch now opens the only remaining current path and the circuit returns to the first state when Y changes again.

Relay Faults

Relay faults arise in three regions: operation current faults, where no current passes through the operating coil; mechanical faults, where the moving parts are acted on by the coil's magnetism but do not move sufficiently (or at all) to close the contacts; and contact faults, where the contacts close, but current does not flow through them.

When a relay is thought to be faulty, the first check is the visual check, looking at the relay as it is operated, to see if the contacts are closing. If the relay is sealed in a case, as frequently happens when long life is required, this test cannot be carried out, and the current through the coil or the voltage across it must be checked. If sufficient voltage or current is applied to the coil for contact closure, check the contact resistance (disconnected) with the relay operated. If this is high (1Ω in most cases) a sealed relay must be replaced.

If an open relay can be seen to be operating, but does not 'make' the contact circuit, the trouble is almost certainly due to dirty contacts. Those should be cleaned using a strip of cloth dipped in a switch-cleaning solvent: abrasives should be used only as a last resort if the contacts are badly burned or pitted and then only as a means of keeping the circuit operating until the relay can be replaced.

In a sealed relay, any relay fault requires replacement, but in an open relay mechanical faults may be dealt with to some extent. This mechanical attention should be confined to cleaning and lubricating (preferably with a smear of silicone grease) the rubbing surfaces only; any adjustments of spring tension or clearances will cause the relay to close and open at different coil voltages or currents and may upset the operation of the circuit.

Devices Related to Relays

Uniselectors, impulse counters and solenoids are electromechanical devices closely related to relays. A *solenoid* is strictly speaking any coil whose length is greater than its diameter, but in electronics work the name is usually reserved for coils used to perform mechanical work by pulling a core into the coil when the coil is energised. Typical applications are found in vending machines, car overdrive units and automatic gear boxes, and various industrial applications, particularly in machinery which automatically rejects faulty items which are then pushed out of the way by the operation of the solenoid.

When a solenoid is coupled to a ratchet which derives a toothed wheel, the mechanism can be arranged so that the energised solenoid pulls the ratchet over a tooth and allows it to engage. When the solenoid is de-energised, a spring acting on the ratchet pulls the wheel round by one tooth. In use, the wheel drives a rotary switch which can carry a very large number of contacts, as many as the wheel has teeth. Each contact is 'made' with a central wiper each time the solenoid is de-energised.

Such a device is the *uniselector* (see Fig. 11.5) which is used when a large number of circuits must be switched in sequence. A peculiar advantage of the uniselector is that it can be 'motored'; the ratchet can be made to open a pair of contacts which are in series with the solenoid. This causes the solenoid to be de-energised, the ratchet returns, pulling the wheel round by one tooth and closing the contacts again so that the process repeats for as long as power is applied.

Fig. 11.5 Uniselector (*Photo courtesy of S.T.C. Ltd.*)

Impulse counters, or counting relays, use a similar driving mechanism to operate a series of make contacts by cams. In this case there is no common contact, and separate circuits can be switched. In many examples, a second solenoid is included which returns the whole bank of contacts to the all-open condition.

Thermal Delay Switches

A problem which is common in high voltage valve circuits is that of ensuring that no high voltage is applied to the valves before the heaters are at their full operating temperature and the bias voltages applied. The output valves of large transmitters frequently require heater currents of 100A or so; if the full heater voltage is applied at once, the current flowing through the cold heater may be 1,000A or more, and the heater voltage must be applied in steps. In both these cases, the sequence of switching could be carried out by manual operation but it is preferable, however, to keep to a single switch and let the later switching be carried out automatically; this can be achieved by thermal delay switches.

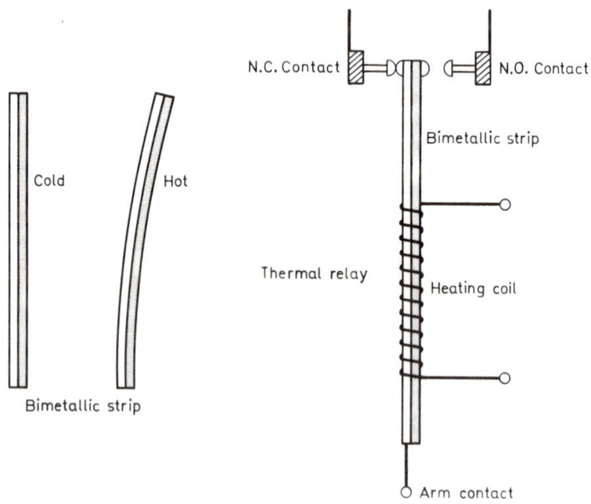

Fig. 11.6 Principle of Thermal Delay Switch.

The operating principle of a TDS is shown in Fig. 11.6. The switching area is made of a bimetallic strip, which is a sandwich of two metals with different expansions, so that one metal strip becomes longer than the other when both are heated. The increase in length is very small, but is sufficient to cause the bimetallic strip to bend noticably when it is heated; the direction of bending is such that the metal with higher expansion is on the outside of the arc of the circle. This bending can be made to close, open or changeover a set of contacts, and the heating coil wrapped round the strip.

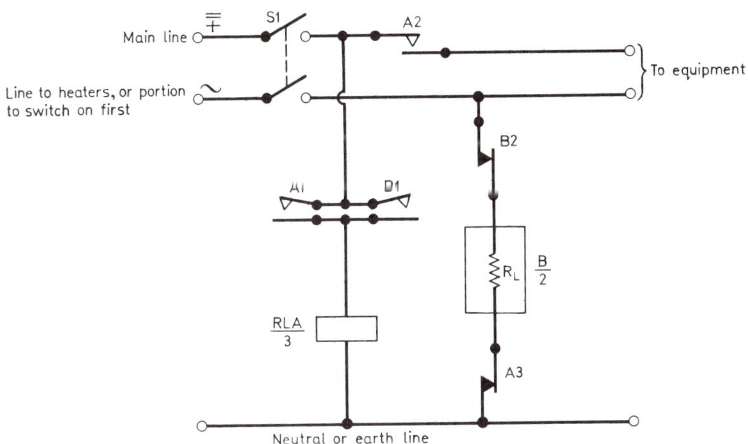

Fig. 11.7 Circuit using thermal delay switch. Switching on S1 applies power at once to the lower main line, which may be a heater supply, and also energises the TDS, RLB. After some time, the TDS operates, closing B1 and opening B2. This operates RLA, closing A1 and A2, and opening A3. As the TDS cools, B1 opens and B2 closes, but the TDS remains cool because of the o/c at A3. If the supply is interrupted, or S1 opened, the circuit returns to its first state when RLA turns off.

Since the heater must be insulated from the bimetal strip (though low voltage, high current switches pass the heating current through the strip itself), the heat takes some time to transfer and the temperature rise of the strip is gradual. At some definite temperature, the strip will bend sufficiently to cause the contacts to switch (on, off, or c/o according to the arrangement).

It would be possible to use the thermal switch permanently in circuit in this way, but this has the disadvantage of keeping the switch hot, and of making it difficult to have the same delay again without a waiting period. For example, should the equipment be switched off for some rèason, and then switched on again, the thermal switch may not have cooled sufficiently to have recovered to its cold position, so that there would be no thermal delay for the second switch-on.

For this reason, circuits of the type shown in Fig. 11.7. are used, so that the TDS is shorted out by a relay once it has made the circuit. The thermal switch can then cool while the equipment is running, and the relay carries the current. If the equipment is switched off, or if there is any other interruption of the current, the relay opens putting the cold TDS into circuit again. A refinement of this method is to use relays to switch alternately two TDSs so that a cold start is always obtainable no matter how many switchings are made.

Thermostats

Thermostats are designed on the same principles as TDSs but the heating of the bimetal strip (or gas, or liquid according to the type) is carried out by the air in the space in which the thermostat is installed. All thermostats have an 'operating differential', meaning that the temperature at which the thermostat switches when the temperature is rising is not the same but higher than the temperature at which it switches back when the temperature is falling. In some types of thermostat, this differential is very low, but it tends to be high in bimetal thermostats, and an accelerator is often fitted to reduce the effect.

The accelerator is a 1W resistor which passes current when the temperature is low (below switching temperature) so that the thermostat switches at a slightly lower air temperature than would otherwise be the case. The switchback temperature is unaffected, because the accelerator heater is shorted by the thermostat contacts until switchback has occurred. If the value of the accelerator resistor is carefully chosen in relation to the position of the thermostat and the size of the space, this system can work quite well.

Microswitches

The name here does not necessarily imply that the whole switch is of very small size, but that the amount of movement required and the force required to operate the switch are very small. Most microswitches use the principle shown in Fig. 11.8. The spring

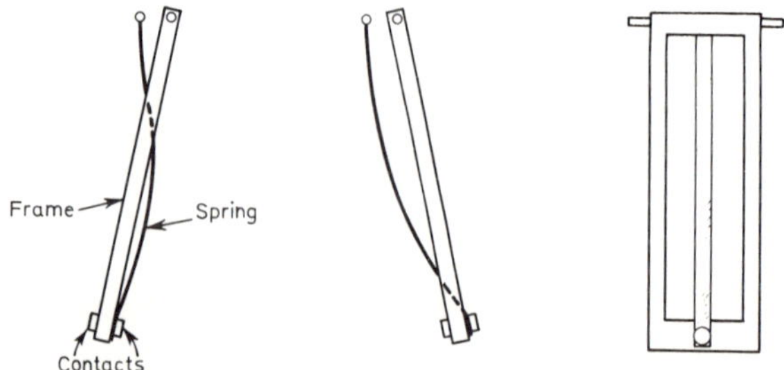

Fig. 11.8 Principle of the microswitch. The spring can hold the switching frame in either of two positions, and a small movement of the frame is enough to make the spring alter its curvature enough to snap the frame over to the other position.

is stable only with the switch open or closed, and it moves the contact rapidly from one position to the other when the force applied to the other end of the spring is altered.

Microswitches are commonly used where a mechanical movement must be limited or reversed. Switches of similar design may be used along with diaphragms for pressure switching in gases and liquids or in balances to indicate that a weight has exceeded limits.

Reed Relays

Reed and diaphragm relays are modern approaches to the problem of relay design in which the contacts and the operating coil have been separated. In reed switches, two or more thin strips of metal are sealed into a glass tube (see Fig. 11.9). The metal is chosen for its

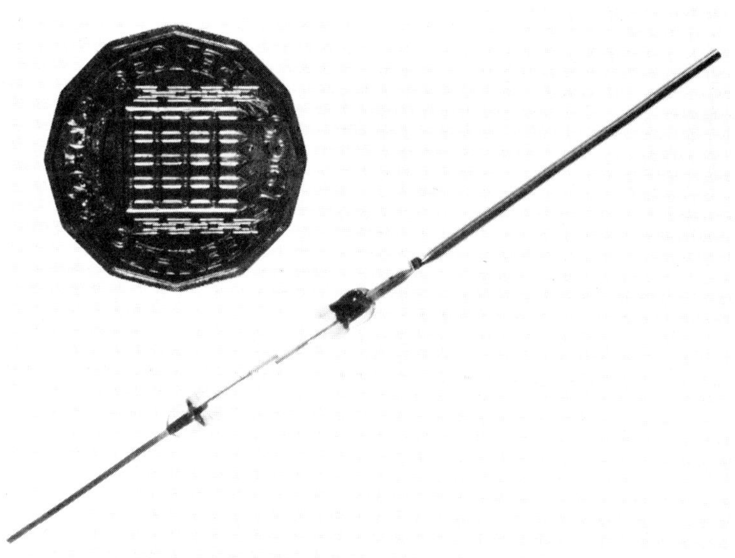

Fig. 11.9 Reed switch (*Photo courtesy of G.E.C. Ltd.*)

springiness and its ability to be easily magnetised, and contacts are formed on the portions of those strips, or reeds, which overlap. When a magnetic field is applied to this reed switch, either from a permanent magnet or from a coil, the reeds are magnetised and come together, making a circuit.

Advantages of Reed Switches

(1) The operating coil can be of low power.

(2) The switch is very small (see Fig. 11.9).

(3) The life of the contacts is very long, because they are completely sealed. In 'sealed relays', the contacts share the seal with the coil. Typical reed minimum life is 200 million operations.

(4) Speed of operation is high, 0·6–1mS.

(5) Reed switches are less likely to be affected by shock or vibration, because the switches are small and light.

(6) Because the reeds are sealed into glass, the insulation resistance between open reeds is very high, around one million MΩ.

(7) Capacitance between open contacts is small, about 0·2pF to 0·8pF. This makes reeds very suitable for switching high frequency signals.

Diaphragm Relays

Diaphragm relays use a thin circular sheet (diaphragm) of metal as the moving contact. The fixed contact is spaced by a film of glass from the moving contact, and carries a rod round which a coil can be placed to magnetise the whole of the fixed contact. The film of glass between the fixed and the moving contacts acts as a seal, and a cover is placed over the moving diaphragm for mechanical protection.

The diaphragm relay has most of the advantage of the reed relay and, in addition, has much larger contact surfaces (so that higher currents can be carried) and lower power operating coils (because of the rod attached to the fixed contact, which makes a more efficient magnetic core).

Contact Bounce

In any device where contacts are closed and opened, problems of contact bounce and contact burning arise. Contact bounce describes the process which takes place when contacts which are spring loaded in some way (as in a relay or a microswitch) make, bounce, break and make again, sometimes to bounce again. In many cases, as when the relay or switch controls a motor or other slow-starting piece of equipment, this does not greatly matter, except that a spark usually passes when the contacts bounce open, causing burning of the contacts.

If the contacts are being used to generate a trigger pulse, however, or if an item of electronic equipment is being 'keyed' (a power supply, an r.f. circuit etc.) then contact bounce can cause considerable trouble (multiple triggers, voltage surges, r.f. pulses) and must be avoided. When the closing of the contacts is used to generate a trigger pulse, it is possible to use the trigger pulse generated by the first closing to generate a gate which prevents subsequent pulses from the 'bounce' from passing to the circuits to be triggered.

In many cases, however, the switch used for generating the trigger is a non-bounce type using a weak spring, and this approach is used when electronic methods of dealing with bounce cannot be applied. Contact bounce on relays is not serious if the relay is operated by the minimum voltage or current necessary to close the contacts, and the design of contact springs on many modern relays has greatly reduced bounce even with excessive operating voltage or currents.

Contact Burning

Contact burning is caused by sparking between the contacts of the relay, mainly when the contacts open a supply to an inductive load. As we have seen earlier, when the current through an inductor is broken, the voltage across the broken portion of the circuit rises momentarily; the voltage peak depends on the current and the rate at which it is broken.

Material	Advantages	Disadvantages	Uses
Silver	low resistance	corrodes	high currents, high contact pressures
Palladium-silver	less easily contaminated	higher resistance	general
Silver-nickel	resists burning resists sticking	higher resistance	general
Tungsten	hard, resists burning	oxidises easily	high power switching
Platinum	very stable to chemical attack	for high voltages and low currents only	teleprinters
Gold	corrosion resistant	low current only	transistor circuits

Table 11.1—MATERIALS FOR RELAY CONTACTS

When this voltage peak (typically several hundred volts when a 12V circuit is broken) exists across the contacts of a relay, sparking takes place, causing contact burning and raising the resistance of the contacts. Eventually this causes pitting of one contact and spikes of metal on the other; this process can be seen on the contact breaker points of a car. Sparking can be reduced by connecting a suitable circuit across the contacts, but care must be taken to avoid connecting a circuit which will effectively lengthen the switching time.

One method is to connect a capacitor and resistor in series across the contacts so that this forms a series tuned circuit when the contacts are opened. The resistor acts as damping to lower the Q of the circuit (preventing free oscillations) and dissipates the power of the inductive circuit, so preventing too large a voltage rise. For this method to be successful, the values of the resistor and the capacitor must be matched to the inductance of the load.

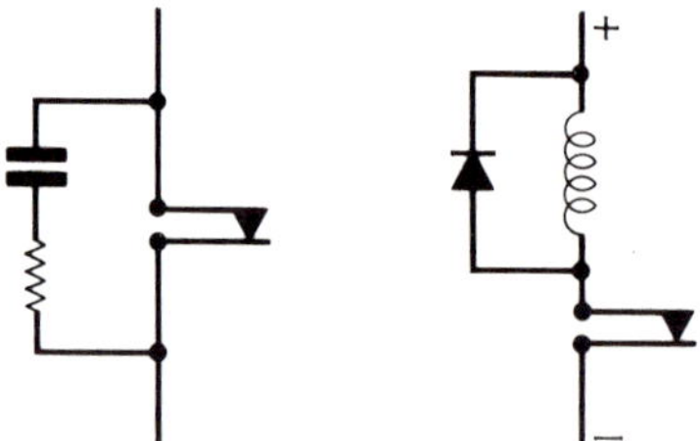

Fig. 11.10 Surge-absorbing circuits.

Another method is to prevent the voltage across the contacts rising above some set voltage, such as line voltage, by the use of a diode as shown in Fig. 11.10. When the contacts are broken, the voltage rise is clipped by the diode which then transfers the energy of the inductive load back to the line until the voltage drops again. The diode must obviously be capable of handling the peak current generated.

Driving Relays and Uniselectors

Since relays, impulse counters, uniselectors etc. are themselves inductive loads to the circuit which drives them, the methods of avoiding contact burn outlined above also apply to preventing damage, particularly to transistors, caused when a relay is switched off. The circuits of Fig. 11.10 are equally applicable to this case.

Magnetic Clutches

Magnetic clutches are used wherever mechanical turning forces must be switched on and off rapidly. Electric motors have heavy moving parts (with the exception of printed circuit motors) and cannot be started and stopped easily. For this reason, when intermittent drives are required, the motors are kept running and the drive is through a clutch which is operated electromagnetically. Basically two forms of clutch are used.

One type uses clutch plates similar in form to these used in the clutch of a car. One plate is fixed to a shaft, another is splined to a second shaft and is free to slide on the splines. Any magnetic force between the plates will cause them to contact together (if they are made of magnetisable material) and transmit the drive from one shaft to the other. The magnetic field can be applied from electromagnets arranged round the plates though not attached to them or rotating with them.

The other basic type, more suitable for high speed operation, is the magnetic powder clutch. Two interlocking cups form the 'plates', one attached to each shaft, and the space inside the cups is filled with the powder of a magnetisable metal. When no magnetic field is present, this powder behaves as a liquid, and very little drive is transmitted. When the powder is magnetised by a coil round the clutch it behaves as a solid piece of material attached to both cups and transmits the drive efficiently. This magnetising can be carried out very rapidly.

Electronic Transducers

A transducer is any device which converts one form of signal into another. A loudspeaker is a transducer which converts electronic signals into sound signals and a gramophone pickup converts mechanical signals (the record track) into electronic signals. When we talk of electronic transducers we are concerned with the conversion of other signals into and from electric signals. The transducers which convert from the electronic signal to other signals (loudspeakers, lamps, heaters, motors) are familiar, but those which convert a variety of signals into electronic signals may be less familiar.

Active transducers generate an electrical output for some other input and need no power supply. Examples are crystals of barium titanate which generate an output proportional to the force acting on them and are used to measure acceleration, selenium cells which give an output proportional to the light falling on them, and

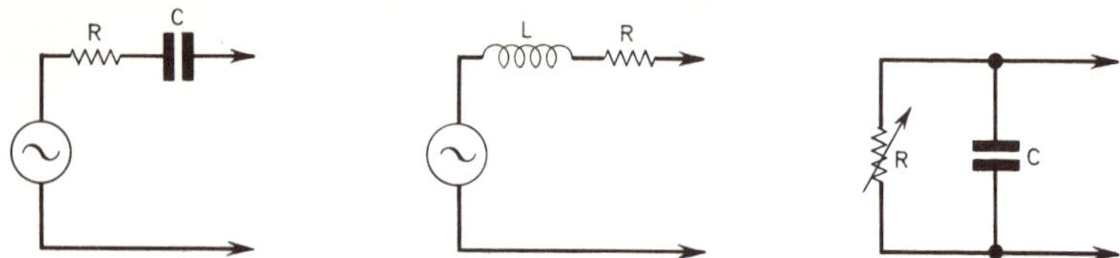

Fig. 11.11 (left) Equivalent circuit of crystal transducer. Typical values of R and C would be 1 MΩ and 2nF.
Fig. 11.12 (centre) Equivalent circuit of inductive transducer. Typical values of L and C might be 10mH and 35Ω.
Fig. 11.13 (right) Equivalent circuit of cadmium sulphide cell. Typical values would be R = 2Ω illuminated, 100 kΩ in darkness, C = 5nF.

thermocouples which give an output proportional to the temperature difference of two connected thermocouples.

Passive transducers act on a 'signal', which may be a d.c. supply or an a.c. signal, in a way controlled by the signal which they are converting into electronic form. Two plates acting as the capacitor in a resonant circuit will change the frequency of resonance if the spacing between the plates is altered—a block of cadmium sulphide alters in resistance as the light which falls on it is varied; a thermistor changes in resistance as its temperature is varied.

Equivalent Circuits of Transducers

Transducers of all kinds usually have one of three possible equivalent circuits. The capacitive and capacitive-resistive equivalent is typified by the equivalent circuit of a crystal pickup and is shown in Fig. 11.11. This type of equivalent circuit is found in acceleration transducers, pressure transducers, force transducers and in all cases where the transducer consists of an insulating material. When the transducer is active, as with a crystal, the output has to be fed into an amplifier whose input time constant with the equivalent circuit capacitance is very long. When the transducer is passive, the transducer usually forms part of a resonant circuit or a capacitive bridge.

The inductive equivalent circuit (Fig. 11.12) is typified by the moving coil microphone and is found in transducers for measuring shaft position or speed or rotation, velocity or rate of liquid flow (where a paddle causes signals to be induced in a coil). Inductive transducers are usually active, and can be coupled directly into an amplifier.

The resistive equivalent circuit (Fig. 11.13) is typified by the cadmium sulphide lightmeter cell, and is found in transducers for measuring mechanical strain (which stretches a filament of the material and alters its resistance), gas pressure and heat transfer. This is usually a passive transducer and the output may often be very small, requiring balanced d.c. amplifiers or chopper amplifiers.

Servomechanisms

A servomechanism is any device which controls the position or speed of an output shaft in response to an electrical signal which may be obtained from the position or speed of another shaft. Power steering in a car is an example of a simple mechanical servomechanism—the effort of steering the front car wheels is carried out by a 'motor' which is directed by the position of the steering wheel—the driver's effort is amplified by the system so that steering can be made effortless even when the forces required are very large, as when the car is stationary.

Synchros are one very common type of servomechanical transducer; Fig. 11.14 shows a synchro control transmitter and receiver.

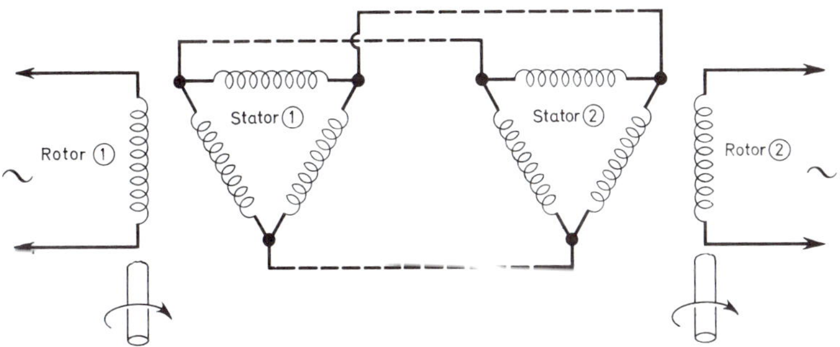

Fig. 11.14 Synchro transmitter and receiver. If both rotors are fed with a.c. of the same phase, movement of one rotor will cause the other rotor to move to the same angular position.

In appearance, servos resemble electric motors, and have also rotor windings and stator windings. In use, the rotor windings (which are turned by the shaft) are supplied with a.c., often of a frequency higher than the 50Hz mains; 500Hz and 1,000Hz being common. Imagine the rotor of a servo transmitter in some fixed position. The servo will now behave as a transformer, and a voltage will be induced in each of the three stator coils, the amplitude and phase of the voltage in each coil depending on the position of the rotor coil.

If the stator coils are connected to the corresponding coils of another identical synchro, the voltages appearing will create magnetic fields which will cause the rotor to turn until the shaft is in the same position as that of the transmitter. If the transmitter synchro shaft is turned, the receiver synchro will follow. The only link between the two is the wiring. Synchro differential transmitters carry three-winding stators and rotors. Rotation of the shaft of a differential transmitter introduces phase difference between the transmitter shaft and the receiver shaft.

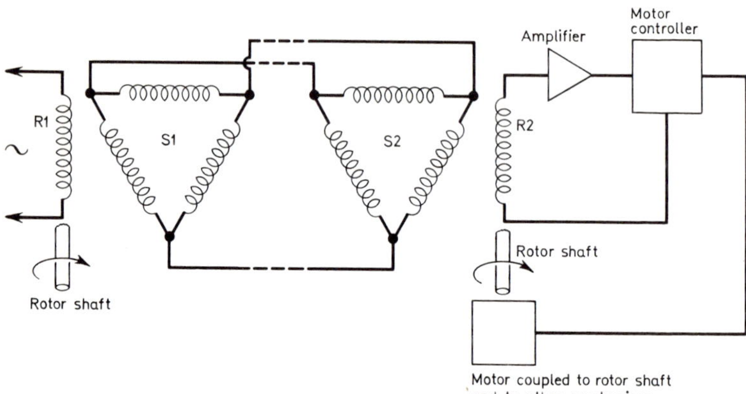

Fig. 11.15 Synchros used in feedback servo system. If rotor (2) does not follow rotor (1), a signal exists across the rotor winding which is amplified and used to turn the rotor shaft until the signal is zero. The motor may also drive other mechanisms. Much greater loads can be driven in this way.

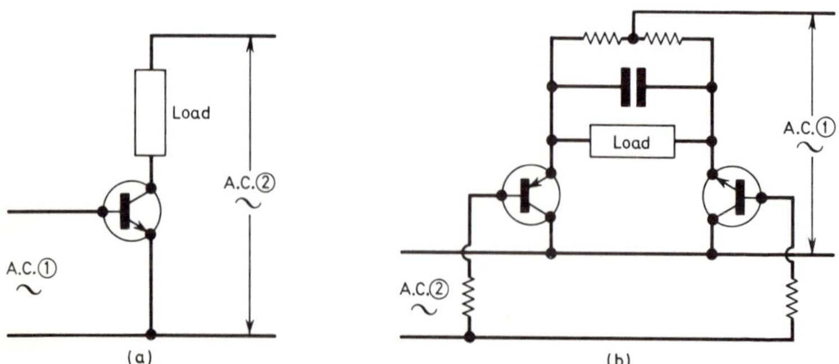

Fig. 11.16 Phase controlled rectifiers: (a) basic circuit. Current in the load is maximum when the two waveforms are in phase. (b) Phase controlled rectifier for small loads. Circuits using thyristors or thyratrons can be used for large power outputs.

In this way, the receiver shaft speed may also be made to be the sum or the difference of the shaft speeds of transmitter and differential.

Synchros may also be used as part of a feedback system. Suppose a transmitter synchro is wired as in Fig. 11.15. If the shaft of the transmitter synchro is moved, the phase of the signal at the rotor winding of the receiver synchro will change. It is possible to generate a signal proportionate to the phase difference which can be amplified and used to control a motor which then turns the shaft of the receiver synchro.

Used in this way, the synchros transmit no power, but are used to signal any error in position, so that the amount of power controlled can be as large as can be provided by a controllable motor. The circuit which delivers an output which depends on the phase of the signals it receives is known as a phase-sensitive rectifier. Two types of phase sensitive rectifiers are shown in Fig. 11.16.

Other Servo Components

Tacho generators are miniature alternators which give an output whose voltage is precisely proportional to the shaft speed of the

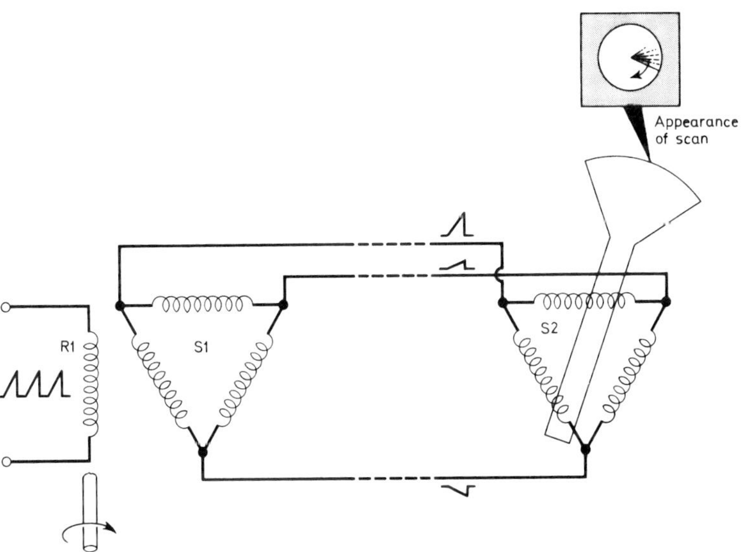

Fig. 11.17 Sweep resolver. R1 is fed with a timebase waveform at low impedance and is rotated by the aerial shaft. S2 is the c.r.t. deflection coil; it produces a display consisting of a line (due to the timebase sawtooth) rotating in step with the aerial.

generator. Most tacho generators require an input (or excitation) supply.

Servo motors are designed to form part of the feedback servo system outlined earlier. They must start moving with the least delay when power is applied (high torque, low inertia) and should be able to turn smoothly even when almost stalled. Servo motors are always a.c. operated, frequently two or three phase, and sometimes have inbuilt tacho generators to form part of the feedback loop.

Synchro sweep resolvers are shaft position transmitters used in radar (Fig. 11.17). When the aerial is rotated by a motor, the shaft of a synchro is rotated by the same shaft. The rotor is fed with a timebase waveform and the output at the stator windings is of this waveform of various amplitudes depending on the position of the rotor. If those waveforms are then passed to three-phase coils used as the deflection coils of a c.r.t. the sweep will consist of a radial line which will rotate round the centre of the tube in step with the rotation of the aerial. Because of the waveform used in the resolver, it must be capable of working at much higher frequencies than those encountered in the servo systems outlined above. Furthermore, the accuracy of bearing angles read on the radar screen depends greatly on the accuracy of the resolver.

An alternative method of achieving the same scan is to rotate a deflection coil fed with a timebase, using a servo system to keep the rotation of the coil in step with the rotation of the aerial scanner.

INDEX

D

Q

R